# GPS
## and Amateur Radio

Walter Fields, W4WCF

**Editor**
Walter Banzhaf, WB1ANE

**Production**
Michelle Bloom, WB1ENT
Sue Fagan, KB1OKW — Front Cover
Jodi Morin, KA1JPA
David Pingree, N1NAS

**ARRL** The national association for AMATEUR RADIO
225 Main Street, Newington, CT 06111-14

# Table of Contents

APRS® is a registered trademark of Bob Bruninga, WB4APR.

# Foreword

The Global Positioning System began as a navigational network for the US military and it still performs that function today. Since the early 1990s, however, it has become an integral part of the non-military world. The GPS user community now includes everyone from pilots to truck drivers. GPS units are becoming commonplace in automobiles; you'll find them on the list of manufacturer options along with cruise control, satellite radio and more.

Of course, Amateur Radio operators have been exploiting GPS almost from the first day it become available to civilians. Its most popular application is in the Automatic Position Reporting System, or APRS. By connecting inexpensive GPS receivers to packet radio modems, hams transmit their positions and other information, which is then displayed on computer-generated maps at the receiving stations. APRS is a fascinating and fun aspect of our diverse hobby, but it also has a serious side in public service applications. For example, APRS has played a prominent role in tracking search and rescue teams.

GPS and Amateur Radio provides a detailed overview of the Global Positioning System, along with advice on how to get the most from your GPS experience. In addition, you'll also find helpful information about ham applications of GPS, including APRS. I trust you'll find this book to be a worthwhile addition to your Amateur Radio library!

David Sumner, K1ZZ
ARRL Chief Executive Officer
March 2007

# Introduction

The marvels of the electronic revolution have completely changed the way we live our lives, and the profusion of electronic devices have become so commonplace that we expect to find new, better, and more efficient electronic devices arriving in the marketplace almost weekly. Few realize that this revolution had its beginning over 100 years ago on November 16, 1904 when British engineer John Ambrose Fleming patented the first electronic device, a thermionic diode or vacuum tube that converted an alternating current into a direct current. This was followed in 1906 by Lee DeForest's invention of the audion tube, a three-element vacuum tube which generated amplification of an electronic signal and made modern radio possible. Many other scientists and engineers continued to contribute to the electronic revolution until in 1948 William Shockley, Walter Brattain, and John Bardeen, working at the Bell Telephone Laboratory, invented the transistor. From that moment on the revolution began to move at exponential speed. Single transistors evolved into multiple transistors arranged on a single silicon chip and connected internally. Single chips evolved into one densely populated integrated circuit capable of performing complex computational and analytical tasks.

The integrated circuit enabled the development of what undoubtedly has become the single electronic innovation that has completely changed the world's view of itself, and added enormous fuel to the explosive electronic revolution. This innovation is the Global Positioning System (GPS). GPS has worked its way into all our lives and made location awareness a common daily activity. It is on our boats, in our cars, in cell phones, in our personal computers, in our PDAs, and Suunto even produces a watch that contains a GPS along with a barometer, an electronic compass, and a PC link for downloading routes.

As GPS becomes omnipresent in our society, its uses in electronic devices continue to grow and spread into almost every arena of our world, replacing old technology and improving the way we navigate, communicate, and move about on the face of this planet. It most definitely has become the navigation tool of choice for boaters, hikers, fishermen, surveyors, aviators, a wide variety of businessmen, and automotive travelers. Position or location awareness has even found its way into Ham Radio with activities like direction finding, APRS, locating the perfect antenna set-up, or finding the right spot for Field Day operation to name a few.

It is the intent of this book to give you the reader an understanding of what the Global Positioning System is all about, its history, how it works, what's needed to use it, how to effectively operate a GPS receiver to accomplish whatever purpose you have in mind, and what to look for in selecting a GPS receiver. Even if you never use GPS to do anything more complicated than find the right location to set up your Field Day operation, once you know how to do that the more useful this system will become, and you will find yourself searching for more things to do with this marvelous electronic tool of American engineering.—*Walter Fields, W4WCF*

# About the Author

Walter Fields, W4WCF, is a retired electronic engineer who participated in the original design of the Global Positioning System, GPS. After graduation from Auburn University with a MSEE he joined the United States Air Force as a Communications-Electronics Officer where he worked in both the Air Force Systems Command and the Air Force Communications Command. While a member of Air Force Systems Command, he worked on an early spacecraft design called Project 621B to provide continuous, three-dimensional navigation, (latitude, longitude and altitude). This led to a new satellite ranging signal based on pseudorandom noise, which is the exact signal used today by the GPS. While serving in Air Force Communications Command he worked on the world's first automatic digital communications system called AUTODIN. This was a computer controlled, worldwide communications system that supported the Air Force Logistics Command and the Air Force Command and Control structure.

He was first licensed in 1967 as KN4TJM and one year later received his General ticket as K4TJM. He has held the previous calls WB6NQX and KG4ROR. Walter presently holds the call W4WCF and is a member of ARRL. Upon his retirement from the Air Force, he joined the engineering consultant firm of Booz, Allen and Hamilton where he provided technical advice and communications and power designs for the Department of Defense and the Executive Department of the government.

Walter retired to Florida in 1994 and joined the United States Power Squadrons (USPS), the world's largest recreational boater organization with some 55,000 members in 455 Squadrons across the United States and Puerto Rico. The USPS is recognized as being the largest private boating educator in the United States, and the "go to" organization for recreational boating knowledge and training. ARRL and USPS have some similarities, in that both have their beginning in 1914, both are the recognized expert organizations in their field, and both believe in providing education opportunities for their members. ARRL and USPS also enjoy a Memorandum of Understanding (MOU) between the two organizations. Walter has held every office at the local level in USPS and now serves at the national level as a Rear Commander.

Walter's books include the *Marine Amateur Radio*, the B*oatowner's Guide to GMDSS and Marine Radio*, the *USPS GPS Learning Guide*, the *USPS Marine Electronics Course*, and the *USPS Marine Engine Maintenance Course*. In addition to the book you're reading now, Walter is presently writing a book to be called *The Boatowner's Guide to Marine Electronics*, which will be published by McGraw-Hill in late 2007.

Walter resides with his wife Ann in Ocala, Florida where they enjoy year round boating activities and Amateur Radio.

# About the ARRL

The seed for Amateur Radio was planted in the 1890s, when Guglielmo Marconi began his experiments in wireless telegraphy. Soon he was joined by dozens, then hundreds, of others who were enthusiastic about sending and receiving messages through the air—some with a commercial interest, but others solely out of a love for this new communications medium. The United States government began licensing Amateur Radio operators in 1912.

By 1914, there were thousands of Amateur Radio operators—hams—in the United States. Hiram Percy Maxim, a leading Hartford, Connecticut inventor and industrialist, saw the need for an organization to band together this fledgling group of radio experimenters. In May 1914 he founded the American Radio Relay League (ARRL) to meet that need.

Today ARRL, with approximately 150,000 members, is the largest organization of radio amateurs in the United States. The ARRL is a not-for-profit organization that:

♦ promotes interest in Amateur Radio communications and experimentation

♦ represents US radio amateurs in legislative matters, and

♦ maintains fraternalism and a high standard of conduct among Amateur Radio operators.

At ARRL headquarters in the Hartford suburb of Newington, the staff helps serve the needs of members. ARRL is also International Secretariat for the International Amateur Radio Union, which is made up of similar societies in 150 countries around the world.

ARRL publishes the monthly journal *QST*, as well as newsletters and many publications covering all aspects of Amateur Radio. Its headquarters station, W1AW, transmits bulletins of interest to radio amateurs and Morse code practice sessions. The ARRL also coordinates an extensive field organization, which includes volunteers who provide technical information and other support services for radio amateurs as well as communications for public-service activities. In addition, ARRL represents US amateurs with the Federal Communications Commission and other government agencies in the US and abroad.

Membership in ARRL means much more than receiving *QST* each month. In addition to the services already described, ARRL offers membership services on a personal level, such as the ARRL Volunteer Examiner Coordinator Program and a QSL bureau.

Full ARRL membership (available only to licensed radio amateurs) gives you a voice in how the affairs of the organization are governed. ARRL policy is set by a Board of Directors (one from each of 15 Divisions). Each year, one-third of the ARRL Board of Directors stands for election by the full members they represent. The day-to-day operation of ARRL HQ is managed by a Chief Executive Officer.

No matter what aspect of Amateur Radio attracts you, ARRL membership is relevant and important. There would be no Amateur Radio as we know it today were it not for the ARRL. We would be happy to welcome you as a member! (An Amateur Radio license is not required for Associate Membership.) For more information about ARRL and answers to any questions you may have about Amateur Radio, write or call:

ARRL—The national association for Amateur Radio
225 Main Street
Newington CT  06111-1494
Voice:  860-594-0200
Fax:  860-594-0259
E-mail:  **hq@arrl.org**
Internet:  **www.arrl.org/**

Prospective new amateurs call (toll-free):
**800-32-NEW HAM** (800-326-3942)
You can also contact us via e-mail at **newham@arrl.org**

# 1 | GPS Basics

## IN THE BEGINNING

The Global Positioning System (GPS) is a satellite-based, radio navigation system that uses 24 orbiting satellites to provide a highly accurate position finding capability to a user of the system anywhere on the face of the Earth anytime, day or night, 24/7. Although GPS has become the best known electronic navigation system today, it was not the first. GPS was preceded by other well known electronic navigational aids including radio direction finders (RDF), hyperbolic systems (OMEGA, DECCA. Loran-A, and Loran-C), and the very first satellite-based navigational aid, TRANSIT.

The evolution of electronic navigational aids from simple radio direction finders to hyperbolic systems to satellite-based systems was a result of a combination of military and other requirements for greater accuracy and technological developments. Electronic navigational systems operate on the principle that electronic signals (radio waves) travel in straight lines (the basis of RDF), reflect from surfaces (the basis of radar), and travel at approximately the speed of light (the basis of hyperbolic and space systems). Distance can therefore be determined by multiplying the signal travel time by the speed of light. Throughout the development of electronic navigation systems this principle has remained the same.

The Global Positioning System

Figure 1.1 – Early Block I GPS satellite. This is an example of the very first GPS satellite.

is owned and managed by the US Department of Defense. The official name of the system is NAVSTAR, which is an acronym for **NAV**igation **S**atellite **T**iming and **R**anging. To meet U S requirements for a highly accurate electronic navigational system for the US military and intelligence communities, the Department of Defense began research and development of GPS in 1973. The United States Air Force was named as the lead agency for this multiservice program. The first GPS satellite, a Block I development model was launched on 22 February 1978. Figure 1.1 is a picture of an early Block I GPS satellite.

Full-scale engineering development of the Block II GPS satellite was completed in 1984, and the first operational Block II satellite was launched on 14 February 1989. Figure 1.2 is a picture of a Block IIA satellite. It wasn't until June 1993, however, that the full constellation of 24 satellites was achieved. Although the system was in use by the military services almost from the launch of the first Block II satellite, it wasn't until after extensive testing of the full constellation of 24 satellites and confirmation of their operational capability, that the Air Force declared the GPS fully operational on July 17, 1995. Figure 1.3 shows a conceptual drawing of the GPS constellation.

GPS was originally developed by the military strictly for military use,

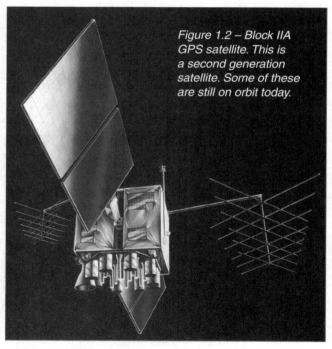

*Figure 1.2 – Block IIA GPS satellite. This is a second generation satellite. Some of these are still on orbit today.*

with no application for civilian use of the system. This changed in 1983 after the downing of Korean Air Flight 007 by the Soviet Union. This tragedy occurred in part because the crew of the Korean 747 aircraft made an error in navigation, which brought the aircraft over Soviet air space. As a result, the Russians shot them down. It was argued that if GPS had been available this tragedy would not have occurred. In the aftermath, President Ronald Reagan issued an Executive Decree that certain portions of the GPS system be made available for civilian use free of charge to the entire world. The US military insisted, however, that those portions of the GPS made available for civilian use be degraded in accuracy so that the system could not be used by the enemies of the US for clandestine purposes. When the Standard Positioning Service portions of the GPS were opened up to civilian use it came with something called Selective Availability (SA) which degraded the normal accuracy of 50 feet to 300 feet. Even with portions of GPS now open to civilian use, there were very few GPS receivers available, and any to be found were very expensive. In 1991, during Operation Desert Storm, the use of GPS was so widespread that the military found they did not have enough GPS receivers to supply the troops. A large multi-sourced procurement by the military for GPS receivers to use in Desert Storm resulted in a tremendous spin-off of the technology into the civilian sector. This, in turn, resulted in the

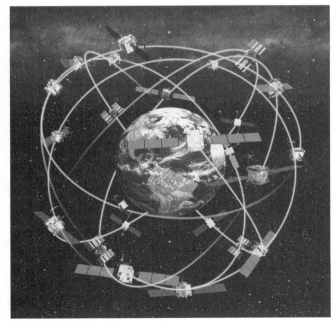

*Figure 1.3 –*
*Conceptual*
*drawing of*
*the GPS*
*constellation.*

availability of highly capable GPS receiver equipment to the civilian market. Although consumer GPS receivers were expensive when they first came on the market, widespread acceptance of the technology and a flood of receiver equipment has resulted today in a basic unit that can give position location accuracy to within 10 feet and can be purchased for less than $100.00.

After many studies and considerable lobbying in Congress, President Clinton ordered that SA be permanently turned off on May 2, 2000. The improvement in GPS accuracy for the civilian world since then has been considerable, and the military has found a way of locally degrading GPS accuracy for selected areas without affecting the rest of the world.

## HOW GPS WORKS

The GPS receiver needs to know two things if it is going to do its job and pass this wealth of information on to the user. It has to know *where* the GPS satellites are in space (location) and how *far away* each of the satellites is from the GPS receiver (distance). Once these two factors are known, the receiver uses a process similar to the process of triangulation you use when fox hunting or T-Hunting, with one significant difference. Instead of trying to determine direction, the receiver measures the *time* it takes to receive each of the signals from each satellite. Let's look first at how the GPS receiver knows where the satellites are located in space.

## LOCATION

The GPS receiver picks up two kinds of coded information from the satellites. One type of information called *almanac* data contains the approximate positions (location) of the satellites. This data is continuously transmitted by the satellites, and when received by the GPS receiver, is stored in the receiver's memory. Each satellite transmits the complete almanac for the entire satellite constellation so your receiver only needs to receive this data from one satellite in order to collect the needed information. Once the almanac data is stored in the receiver's memory, the receiver knows the individual orbits of the satellites and where they are supposed to be in space. The almanac data is periodically updated with new information as the satellites move around.

Any satellite can travel slightly out of orbit, so ground monitor stations keep track of each satellite's orbit, altitude, location, and speed. The ground monitor stations send this data to a master control station which analyzes the data, makes corrections based upon a highly accurate computer model, and sends the corrected data up to the satellites to update their almanac. The corrected position data is called the *ephemeris*, and is valid for about six hours.

After that time period, it must be corrected again by the master control station. The ephemeris is transmitted by the satellites to GPS receivers as part of a coded system called the C/A or *course acquisition code*.

## DISTANCE

Even though the GPS receiver knows the precise location of each of the satellites in space, it still needs to know how far away the satellites are in order to determine a position on Earth. There is a simple formula that the receiver uses that tells it how far it is from each satellite.

The distance from a given satellite equals the velocity of the transmitted signal from the satellite multiplied by the time it takes the signal to reach the receiver (Distance = Velocity × Travel Time). Let's use an analogy to explain this. Most of us at an early age learned that we could find out how far a thunderstorm was from us by observing a lightning flash and counting the number of seconds until we heard the thunder. The longer the count, the further away the storm was from us. In the case of the thunderstorm, we used the speed of sound at sea level for the velocity (approximately 0.2 miles per second) times the travel time (number of seconds we counted) to determine the distance. GPS works on the same principle, called Time of Travel.

Using the same basic formula to determine distance, the receiver already knows the velocity. It's the speed of a radio wave, approximately the speed of light, less any delay as the radio signal travels through the earth's atmosphere.

Now, the receiver needs to know the time factor of the formula. The answer lies in the coded signals each satellite transmits. The time factor is contained in something called the "pseudo random noise code" or the PRN code because it looks like a noise signal. The PRN code is a series of ones and zeros (binary) 1023 bits long that takes 1 millisecond to transmit. The PRN code contains a great deal of information that the receiver uses to identify each satellite and to determine its position in space. When a satellite is generating the PRN code, the GPS receiver is attempting to generate the same code and tries to match it to the satellite's PRN code by comparing the two codes and determining how much it must shift or delay the code it is generating to match the satellite code. Once the amount of shift or delay (called the pseudo range) is determined, the receiver multiplies it by the velocity (speed of light) to get the distance. See Figure 1.4.

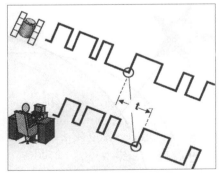

*Figure 1.4 – Example of pseudo range.*

GPS receivers do not keep time as precisely as the satellite clocks, since the satellites' clocks are atomic clocks and the receiver's clock is a quartz clock similar to what is in the ordinary wrist watch. Each distance measurement computed by the GPS receiver, then, must be corrected to account for any internal clock error. For this reason, the range (distance) measurement is referred to as a "pseudo range". To determine position using pseudo range data a minimum of four satellites must be tracked and the four individual position fixes must be recomputed until the clock error disappears.

## DETERMINING POSITION

Now that the receiver has both satellite location and distance, it can determine a position. Let's say the receiver is 11,000 miles from satellite W. The location of the receiver would be somewhere on an imaginary sphere that has satellite W in the center with a radius of 11,000 miles. See Figure 1.5. Let's say the receiver is also 12,000 miles from satellite X. The second sphere would intersect the first sphere to create a circle common to both spheres. See Figure 1.6. Now, if a third satellite, satellite Y, is added at 13,000 miles from the receiver, we have two common points where the three spheres intersect. See Figure 1.7.

We now have two possible solutions for the location of the receiver, but the two solutions differ greatly. One of the solutions determines a position on Earth, but the other determines a position in space. The position in space is obviously incorrect so the receiver discards this solution in favor of the position on Earth. If a fourth satellite was added to this solution the accuracy of the position fix would be greatly increased and the receiver clock would be completely upgraded and in sync with the timing signal of the satellites' atomic clocks. See Figure 1.8, which shows the total GPS solution with 4 satellites acquired and the position fix in latitude, longitude, altitude and time.

The GPS system consists of three segments: the space segment, operated by the United States Air Force (USAF); the control segment, also operated by the USAF, and the user segment (or GPS receiver). Let's look at a little more detail of each of these three parts of the GPS.

## SPACE SEGMENT

The space segment consists of 24 GPS orbiting satellites. Twenty-one satellites are active (turned on) and three are on-orbit spares (turned-off). Sometimes there are more than 24 usable GPS satellites on orbit as new satellites are added to the constellation. As we learned earlier, three satellites are required to obtain what is called a two-dimensional GPS fix (latitude and longitude). A three-dimensional fix (latitude, longitude, altitude and

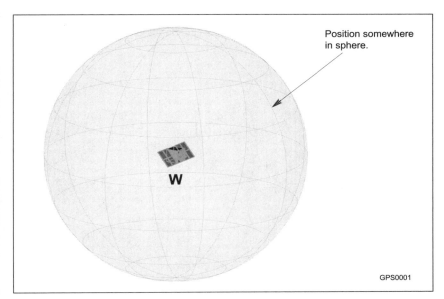

*Figure 1.5 – Example of signal from satellite W locating the receiver on an imaginary sphere in space.*

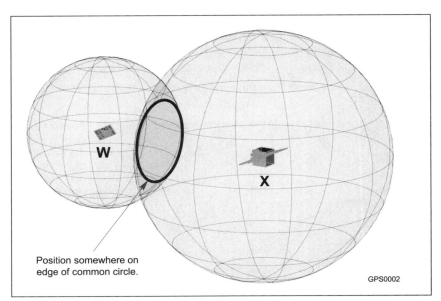

*Figure 1.6 – Example of how adding a 2nd satellite (satellite X) describes a circle of position (COP), defined by the intersection of the two spheres.*

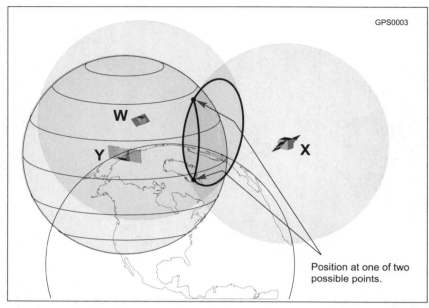

W

Y

X

Position at one of two
possible points.

*Figure 1.7 – Adding a 3rd satellite (satellite Y) further defines two possible
locations for the receiver on the COP and provides a simple position fix.*

time) requires four satellites. The accuracy of the GPS fix is improved using
additional GPS satellites, if visible.

GPS satellites are inserted into six different circular orbit paths 10,898
nautical miles above the Earth. Each of the six orbital planes normally contains
four satellites that are inclined 55° to the Earth's equator.

Satellites are separated by 60° in each plane; therefore a plane could
contain up to six satellites. The orbit period for each GPS satellite is 11 hours

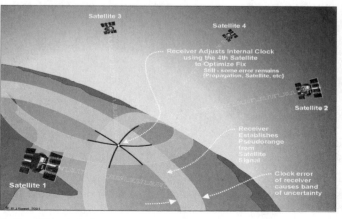

*Figure 1.8 –
Adding the
4th satellite
gives an
accurate
position fix
as well as
altitude and
time.*

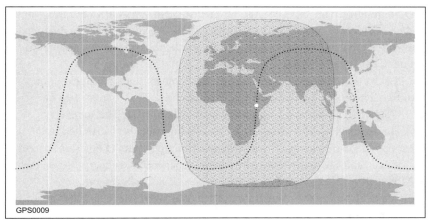

*Figure 1.9 – Diagram detailing the satellite's orbit and ground tracks.*

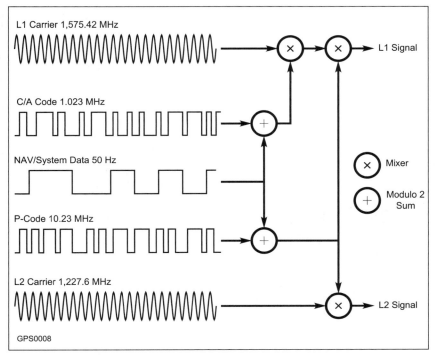

*Figure 1.10 – Diagram showing how the L1 and L2 signals are generated.*

and 58 minutes. This orbit configuration ensures that six GPS satellites will be in line of sight for most viewers any place on Earth. Figure 1.9 is a diagram showing a GPS satellite's orbit around the earth, and its resulting ground track on the Earth's surface.

Each GPS satellite contains two 50-watt transmitters that transmit on two frequencies, called the L1 and L2 frequencies. The L1 frequency is at 1575.42 MHz and the L2 frequency is at 1227.6 MHz. The standard positioning service (SPS) signal used by civilian users is transmitted in the clear on the L1 frequency. The precise positioning system (PPS) signal, used by United States Military and Government departments, is encrypted on the L2 frequency. Figure 1.10 illustrates how the GPS satellite signals are generated. Note that the navigation/system data 50 Hz signal is present in both the L1 (SPS) and L2 (PPS) signals.

The heart of each GPS satellite is its clock. The older GPS satellites (Block II and IIA launched prior to 1997) have cesium atomic clocks. The newer GPS Block IIA and IIR satellites (launched in 1997 and later) have rubidium atomic clocks. Figure 1.11 is a picture of a rubidium atomic clock used on Block IIR satellites. The satellites' atomic clocks keep accurate time to within three nanoseconds; that's 0.000000003 second, or three billionths of a second.

The satellites are identified by the receiver by means of PRN-numbers. Real GPS satellites are numbered from 1 to 32. These PRN-numbers of the satellites appear on the satellite view screens of many GPS receivers. For simplification of the satellite network 32 different PRN-numbers are available, although only 24 satellites were necessary and planned in the beginning. For a couple of years now, more than 24 satellites are available, which optimizes the availability, reliability and accuracy of the network.

The mentioned PRN-codes are only pseudo random. If the codes were actually random, $2^{1023}$ possibilities would exist. Of these many codes, only a few are suitable for the auto correlation or cross correlation which is necessary for the measurement of the signal propagation time. This noise-like code modulates the L1 carrier signal, which is the basis for the standard positioning service (SPS).

Each Block II GPS satellite in orbit weighs approximately 2,000 pounds and measures approximately 17 feet across its solar panels. The design life of a Block II/IIA satellite is 7.5 years, with a mean mission duration of 6 years. However,

Figure 1.11 – A rubidium atomic clock.

several Block II satellites have been operational in orbit for over 12 years. The third generation Block IIR satellites weigh approximately 2,370 pounds. Their solar panels generate 1136 watts of electricity, and the satellite and its solar panel measure over 38 feet. The design life of a Block IIR satellite is 10 years, with mean mission duration of 7.5 years.

## CONTROL SEGMENT

The GPS-System is controlled by the US Air Force. The "master control station" (Schriever AFB) and four additional monitoring stations (on Hawaii, Ascension Islands, Diego Garcia and Kwajalein) were originally set up for monitoring the satellites. During August and September 2005, six more monitoring stations of the NGA (National Geospatial-Intelligence Agency) were added to the grid. Now, every satellite can be seen from at least two monitoring stations. This allows the master control station to calculate more precise orbits and ephemeris data. For the end user, a better position precision can be expected from this. Figure 1.12 shows the GPS Control Segment configuration.

In the near future, five more NGA stations will be added so that every satellite can be seen by at least three monitoring stations. This will improve the integrity monitoring of the satellites and thus the whole system.

The purpose of the Control Segment is to provide corrected orbital data (ephemeris) and corrected clock (time) information to each GPS satellite. The primary master control station is located at Schriever AFB, Colorado (East of Colorado Springs). The back-up control station is located in Gaithersburg, MD. The GPS master control station is operated for the Department of Defense by the US Air Force's 2nd Space Operations Squadron of the 50th Space Wing.

*Figure 1.12 – The GPS control segment configuration showing how the monitor stations are linked to the master control station.*

Figure 1.13 – Aerial view of Schriever AFB, home of the GPS master control station.

Figure 1.13 is an aerial view of Schriever AFB and Headquarters of the 50th Space Wing.

The master control station computes extremely accurate orbits for each of the GPS satellites. These computations are based on information from the GPS monitor stations, located throughout the world, which collect information from all of the GPS satellites in their view and send it to the master control station. In addition, the master control station receives tracking information from other United States Government ground-based optical and ground-based radar tracking stations. The master control station computes precise current location and future location predictions for each GPS satellite. It then formats the data into discrete ephemeris (current and future location predictions) messages for each GPS satellite. The ephemeris updates, and sometimes clock corrections, are then uplinked to each GPS satellite through a number of GPS ground-based radio command links. The GPS radio command links are also used to receive telemetry (satellite control and monitoring) information from each GPS satellite and to confirm the receipt of commands and updates from the GPS master control station. The GPS tracking, telemetry, and control (TT&C) signals are in the S Band with the downlink at 2227.5 MHz and the uplink at 1783.74 MHz.

## USER SEGMENT

The user segment is the GPS receiver that is used to acquire navigational information from the GPS satellites. It is called a GPS receiver but, in reality, it is an *integrated* GPS receiver, a

Figure 1.14 – Photo of a general-purpose GPS handheld receiver.

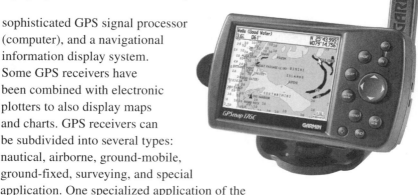

sophisticated GPS signal processor (computer), and a navigational information display system. Some GPS receivers have been combined with electronic plotters to also display maps and charts. GPS receivers can be subdivided into several types: nautical, airborne, ground-mobile, ground-fixed, surveying, and special application. One specialized application of the GPS signal concerns timing where only the time information is used from the GPS satellites. The software in the different types of GPS receivers has been tailored and optimized for that particular application. For example, nautical GPS receivers may not provide accurate navigational information when used in an airplane due to the much faster velocity of the aircraft and changes in altitude above sea level. Figure 1.14 is a typical general purpose handheld GPS receiver, and Figure 1.15 is a GPS receiver and chart plotter.

# 2

# GPS Accuracy

Highly accurate positioning information on a continuous worldwide basis is the hallmark of the Global Positioning System. The GPS has two positioning systems that provide different degrees of accuracy. They are the Precise Positioning System (PPS) and the Standard Positioning Service (SPS). The Precise Positioning System is totally encrypted and can only be accessed by authorized US Government personnel. Obviously, this positioning system is the more accurate of the two as its name implies; however, the exact accuracy of this system is a highly classified secret and unknown outside the US Department of Defense. The Standard Positioning Service is that portion of the GPS that is available to civilian users.

When measuring accuracy of any navigation positioning device, engineers use one or more probability formulas. One that you see quite often is "circular error of probability" (c.e.p.). Another is "distance, root mean squared" (Drms), which is a less optimistic means of determining error than c.e.p. when computing system accuracy. For the Standard Positioning Service the probability of error is computed to be 2Drms. This means that 95 percent of the time there will be no accuracy error with this system. That leaves 5 percent of the time we can expect some error to occur with our GPS receiver. The sources of those errors can include many factors such as:

1. *Ionosphere and troposphere delays.* While in the vacuum of space the radio signals transmitted from the satellites travel at the speed of light; however, once they enter the earth's atmosphere they slow down to something less than the speed of light. Most GPS receivers have a built-in model that calculates an average delay, but not an exact amount. Since the satellite distance factor depends upon signal velocity, even small errors in calculated velocity can throw positioning accuracy off by as much as several meters. The Precise Positioning System is not affected by ionospheric and tropospheric delays since it uses both the L1 and L2 signals to determine satellite distance.

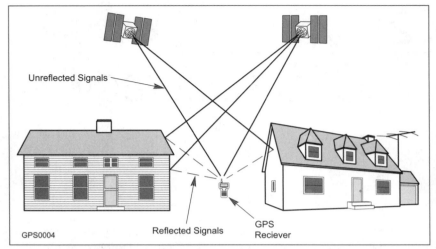

*Figure 2.1 – Illustration showing how the GPS signals can be directed into multiple paths, resulting in deteriorated accuracy.*

    2. *Signal multi-path.* This occurs when the GPS signal is reflected off objects such as tall buildings or large rock surfaces before it reaches the GPS receiver. This increases the travel time of the signal, thus causing delays which can result in errors in position location. See Figure 2.1.

    3. *Receiver clock errors.* The receiver clock is a quartz clock and subject to timing errors. To achieve a position accuracy of 10 meters the run time of the GPS signal must be precise to 0.00000003 seconds, or 30 nanoseconds. This can be achieved only with a quartz clock if the receiver's clock is precisely synchronized with the atomic clock of the satellite.

    4. *Orbital errors.* Also known as "ephemeris errors". These are inaccuracies of the satellite's reported position that have not been corrected by the master control station.

    5. *Number of satellites visible to the receiver.* The more satellites the receiver can "see", the better the accuracy. While three satellites will give a position reading, four satellites will ensure clock errors are reduced. Buildings, terrain, antenna position, electrical interference, or sometimes even dense fog can block signal reception, causing position errors. The clearer the view between the satellites and the receiver, the better the reception.

    6. *Satellite geometry.* This refers to the relative position of the satellites at any given time. Ideal geometry exists when the satellites are located at wide angles relative to each other. Poor geometry results when the satellites are located in a line or in a tight grouping. Figure 2.2 is a graphic illustration of poor and good satellite geometry.

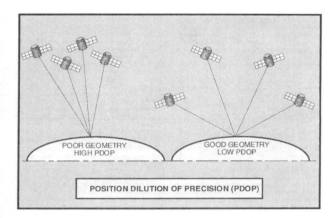

Figure 2.2 –
Diagram showing
bad and good
satellite geometry.

7. *Intentional degradation of the satellite signal.* This occurs when Selective Availability is turned on. At this time this is not a factor as SA is permanently turned off.

So, what is the accuracy of the GPS that we can expect when using this system for navigation purposes? Originally, the system was programmed to be accurate within 100 meters (about 300 feet) 95 percent of the time with SA turned on. Now that SA is turned off we can expect an accuracy of about 30 meters (about 90 feet) 95 percent of the time. When many of those other sources of error are not present, we can expect an accuracy of 6 to 12 meters (about 20 to 40 feet) 95 percent of the time.

However, even this accuracy can be improved by combining the GPS signal at the receiver with one or more of the augmentation systems that have recently come into being and made available for our use. The two augmentation systems most frequently used are the Differential GPS (DGPS) and the Wide Area Augmentation System (WAAS).

## DGPS

When the United States Coast Guard was operated by the US Department of Transportation it did not have access to the Precise Positioning System operated by the Department of Defense. Since 100 meter or even 30 meter accuracy would not satisfy their need for precise harbor entrance and approach, especially in inclement weather, the USCG decided to augment the GPS with their own system called Differential GPS.

DGPS uses 50 remote radiobeacon broadcast sites around the Continental US, Hawaii, and Puerto Rico to provide correction data to the GPS signal and enhance GPS accuracy. Besides the 50 remote sites, there are two control

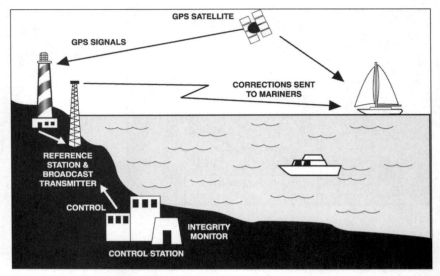

Figure 2.3 – The DGPS configuration.

stations in the system that manage the remote sites and monitor the integrity of the system. The East Coast control station is located at the USCG Navigation Center in Alexandria, Virginia, and the West Coast control station is located at the Navigation Center Detachment in Petaluma, California.

Here's how the system works. At the 50 remote radiobeacon sites, a geodetic survey was performed to determine the precise latitude and longitude of each site. Standard GPS receivers were then installed at these precise locations. When the Standard Positioning signals were received from the GPS satellites at these sites, the positioning data was compared to the exact known position of each site and any required corrections were noted. The correction data was then transmitted on the site's radiobeacon frequency (285 to 325 kHz) to users who had medium frequency receivers connected to their GPS receivers. The GPS receivers then applied the correction data and displayed the corrected positioning information. Figure 2.3 is a drawing showing the configuration of the DGPS.

The system typically increases the GPS accuracy from a nominal 30 meters to 10 meters or less 98 percent of the time. There are some disadvantages to this system, however, in that it requires a separate medium frequency receiver to receive the correction data, and the coverage is limited to the coastal areas of the United States, the Great Lakes, the Mississippi River area, Hawaii and Puerto Rico. It is not a worldwide system as is the GPS.

# THE WIDE AREA AUGMENTATION SYSTEM (WAAS)

When the US Government announced in May 2000 that Selective Availability, that part of the GPS that degraded positioning accuracy, would be permanently turned off, civilian users rejoiced. Using GPS for navigation just got better and at last, electronic navigation had become a reliable and principal source to use and trust. Then, in 2003 the Department of Transportation announced the implementation of an advanced radionavigation system called the Wide Area Augmentation System (WAAS) that augments GPS to provide greater accuracy, system integrity, and system reliability. WAAS was designed primarily to provide navigation and precision approach and landing for aviation users, but has great application for all other users as well. Unlike Differential GPS which broadcasts correction data from land based stations and uses a specialized receiver to provide that correction data to the GPS receiver, the WAAS correction signal is broadcast from space and uses the same signal frequency as that of the GPS Standard Position Service. This means the WAAS area coverage is much greater – roughly from the middle of the Atlantic Ocean to the middle of the Pacific – and does not require additional receivers separate from the GPS receiver.

Here's how it works. The normal GPS satellite data are received and processed at widely dispersed ground based sites called Wide Area Reference Stations (WRS). The received data are forwarded to processing sites referred to as Wide Area Master Stations (WMS) which process the data to determine the integrity, differential corrections, residual errors, and ionospheric information for each monitored satellite. Using the processed data, the WMS generate Geostationary Earth Orbit (GEO) satellite parameters and send this information to a Ground Earth Station (GES). The GES then uplinks the data along

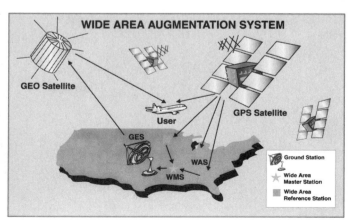

Figure 2.4 – The WAAS configuration.

with the GEO navigation message to the GEO satellites. The GEO satellites downlink the data on the normal GPS L1 frequency with a modulation similar to that used by the GPS. The result is a highly accurate positioning signal that can be used by aircraft for navigation and precision approach and landing. For all others, it means highly accurate positioning information through the use of a single navigation receiver that provides wide area coverage for the 48 contiguous states, Hawaii, Puerto Rico, and Alaska. In fact, the GEO satellites provide coverage halfway around the entire Northern Hemisphere. The horizontal accuracy of WAAS has been measured by the Department of Transportation at 3 meters 95 percent of the time. The makes WAAS the most accurate GPS based navigation system available to civilian users. Figure 2.4 shows the configuration of the WAAS.

## WHAT'S IN THE FUTURE?

In September 2005 the Air Force launched the first of a new series of satellites called GPS IIR-M. This new series will not only broadcast the navigation signals as the previous satellites, but will also provide two new military signals and a second civilian signal. These improvements promise to bring greater accuracy, added resistance to interference, and enhanced performance for users around the world. The second civilian signal removes navigation errors caused by the Earth's ionosphere and increases the civilian accuracy to a few feet rather than a few meters. Once testing of the new GPS IIR-M1 is complete, three more of the new series are planned for launch. However, it will be some time in the future before the full constellation of the 28 planned satellites is complete.

Once the new series is fully operational, civilian users will get better than WAAS accuracy without the need of WAAS, but a new GPS receiver will be needed to take advantage of the improvement in accuracy.

# 3

# GPS Receivers

The GPS receiver is a marvelous navigator whose basic function is to receive and process radio navigation signals from the GPS satellites, and compute a set of coordinates in longitude and latitude for your present position. That all of this can be done with a device that can be held in the palm of one's hand is a tribute to American ingenuity and engineering. In addition to the basic function of positioning information, most GPS receivers have been expanded with additional software/firmware that can be used to supply you with a vast array of useful information. We have come a long way from the first portable GPS receiver that was developed for the military. It weighed about 15 pounds and was designed to be carried in a back pack. The original cost was about $40,000 and it could only receive one satellite at a time. Some of today's GPS receivers weigh about 6 ounces, cost less than $100, and can receive twelve channels of information simultaneously while also computing additional information such as distance to the next waypoint.

In considering the use of the GPS for navigation purposes, we need to first look at the basic principles of navigation. The navigation problem can be broken down into five basic questions:

1. Where am I? This can be answered by your GPS.

2. Where do I want to go? This is answered by you.

3. What course or route should I take? If waypoints are used and the answer to question five is "No," this question can be answered by your GPS.

4. When will I arrive? This can be answered by your GPS, especially if you use enroute waypoints.

5. Are there any hazards enroute (between waypoints) that must be avoided? The navigator can answer this question only after careful examination of the intended route on a map or a nautical chart. GPS does not (cannot) answer this question and automatically pick or suggest a safe route. GPS blindly routes you from waypoint to waypoint with no knowledge of your surroundings.

## USING YOUR GPS RECEIVER

To get the most from your GPS receiver, it is important to read the user's manual that came with it. It is also important that you take some time to familiarize yourself with the operating features of your GPS receiver, and have some understanding of the functions of the buttons on the keyboard. When you first power up your GPS receiver, you have taken the first step in exploring the wonderful possibilities of electronic navigation. At this point you should be aware that the GPS receiver only has a few buttons, but many functions. If you have read your user's manual, you are aware that the receiver uses menus and sub-menus to set up and access the programmed features of the GPS receiver. Understanding the structure of these menus and sub-menus can be a real source of frustration and confusion for the first-time user. In many cases, you will have to work through several layers of menus or sub-menus to get to that one function or operation you wish your GPS receiver to perform. As you use your receiver and become familiar with its operation, the use of these menus and sub-menus will become more apparent. Don't be afraid to push the buttons on the keyboard; you will have to work very hard to break anything. Now, let's go about initializing the receiver so that it can access the satellites, and we can begin our journey into electronic navigation.

When you first turn on your receiver, you will need to be outside in an area that will give you a clear view of the sky. Once you turn on the receiver, it will start to acquire the satellites. The first page you will see after you turn the set on will be a "welcome" page or a manufacturer's disclaimer page. This will not last very long, and the next page will be what we will call the "satellite/status page." You don't have to do anything else at this point; the receiver will eventually acquire the satellites needed to perform its design function. This is called a "cold start," and can take several minutes (about 17-20 minutes depending on the model of receiver). Be patient, it will work. If your receiver has never been used before, or you have previously used it in

another location, the receiver has no idea which satellites are in view when it was turned on, so it must go through the entire almanac data until it finds the satellites that should be in view at the time and place of your location. Once these satellites are found, the receiver begins to process information received from the satellites to establish a fix in latitude and longitude at your location. Unless

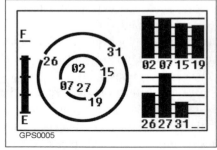

*Figure 3.1 – Satellite page showing positions of seven satellites.*

you move some distance from your present location, the next time you turn the receiver on, it will take only a few seconds for the initialization process to take place. This is called a "warm start." Now that we have initialized our receiver, let's take a look at the "satellite/status page" that is presently in view, and discover some of its features.

Figure 3.1 is a representation of a typical GPS receiver's satellite page. The first thing we should notice is that it provides a visual reference of the current satellite coverage. The heavy black bars on the right side of the display represent the received signal strength of the various satellites the receiver is acquiring. Notice also, that the receiver has identified each satellite by its PRN number. The PRN numbers displayed are being called up from the almanac data in the computer memory of the receiver. These are the satellites the receiver is expecting to see at your position at the specific time carried on the receiver's clock. On the left side of the display there are two concentric circles, one large circle and one small circle with a black dot in the center. The large circle represents the horizon from your position, and the small circle represents 45° above the horizon. The black dot in the center represents a position 90° above the horizon, or directly overhead at your position.

In Figure 3.1, we see that satellites 26 and 31 are right on the horizon. They may be rising or setting. The only way we can tell is to watch the display to see if they disappear. If they disappear, they are setting. In any event, they are not good candidates for obtaining positioning data at the specific time of the display. Satellites 07 and 15 are 45° above the horizon, and are good candidates for the receiver to use. Also satellite 19 is rising to 45° above the horizon, and satellites 02 and 27 are also above 45° from the horizon. These satellites would normally be good candidates for our receiver to use. The only thing wrong with this group of satellites is that satellites 07, 27, and 19 are bunched fairly close together. This could possibly give us a positioning error called geometric dilution of position (GDOP). Some GPS receivers will display on the satellite/status page a GDOP number based on a scale of one to ten. Lower numbers are the best, and higher numbers are the poorest. If your receiver has such a display, then a GDOP number of 5 or greater means the accuracy of the positioning data is probably in error. This could mean that your actual position could be as much as 100 meters or as little as 15 meters off from that indicated on your receiver.

Some GPS receivers have an area on the top left of the display called "coverage." This is where the receiver informs the user of the status of the operation of the receiver. For example, if the receiver is searching for satellites, the term "searching" will appear. If there is interference or obstructions to the receiver's antenna, the term "poor coverage" might appear. Also, if the receiver is a battery-operated model, there will usually be an indication of the battery status on this page. In Figure 3.1, battery status is represented by the heavy black line between "F" for full and "E" for empty. Some GPS receivers will also display additional status information such as "EPE 49 FT." In this case, EPE stands for "Estimated Position Error" and 49 feet represents the error that detracts from the accuracy your GPS believes you should be achieving. Other receiver status that could be displayed, depending upon the make and model of receiver, is as follows:

- **"Looking for Satellites**," reporting no usable GPS signal
- **"AutoLocate**," reporting Cold Start with no current almanac data in memory
- **"Acquiring Satellites**," reporting warm start with current almanac data in memory
- **"2D**," reporting good signal from only three satellites with good geometry
- **"3D** "(described above), reporting good signal from four or more satellites with good geometry
- **"Receiver Not Usable**," reporting that receiver and its integrated computer has failed self diagnostics
- **"Simulation**," reporting that data are simulated and not real

From the discussion above, you can see a wealth of information is available to the user just on this one page. Knowing this information can help you use your GPS receiver successfully.

## Receiver Keyboard Function

Not all GPS receivers have the same set of buttons, and some buttons have different names among different receivers although they might serve the same function. Some receivers have dedicated buttons for some tasks, and other receivers require you to manipulate other buttons to activate that task. Some buttons perform dual functions, with one push of the button to activate the primary function, and two pushes required to activate the secondary function. Of course, this could lead to further frustration for user. To alleviate some of these frustrations, study your user manual. What is presented here is a representative list of those buttons found on most GPS receivers, along with a brief description of the function or task associated with that button.

*Power Button*. This button will turn the unit on and turn it off. One firm push of this button will usually turn the unit on. To turn it off, this button usually must be held down for a few seconds. If this is the case, the screen will display a message that tells you how many seconds you must hold the button down to turn the unit off. Some GPS receivers also use this button to turn on the backlight needed to clearly see the display screen in poor lighting conditions. In this case, turn on the backlight by briefly pressing the power button while the unit is operating.

**Mark Button.** Pressing this button will capture and hold in the receiver's memory your present position in latitude and longitude. It will also bring up a "Mark Position Page" on some GPS receivers that will allow you to give the marked position a name and enter it into your waypoint list. We will discuss the waypoint list in detail later. In any event, the receiver will assign the marked position with a number that you can later change to a name and add to your waypoints. Some receivers do not have a separate "mark" key. In this case, you generally access this function as a secondary feature of a dual function button such as the ENTER button.

**ENTER Button.** This button is analogous to the "ENTER" key on the personal computer (PC), and is used in much the same way. When you press this button, the GPS receiver will execute whatever task you have established for it. For example, when you highlight a task on one of the menu pages and press the ENTER button, that task will be performed. On some receivers, the ENTER button has a secondary function, such as the mark position function, and can be accessed by pressing the button twice.

**GOTO Button.** You use this button to tell the receiver where you wish to go. When you press the GOTO button, it will usually bring up the stored waypoint list. You can then select a waypoint where you wish to go, and the receiver will compute a course to that waypoint.

**Menu Button.** If your receiver has this button, by pressing it, you can have direct access to the menus that control the receiver. If your receiver has more than one menu page, you can access all of the other menu pages by repeatedly pressing this button. To access any of the functions displayed on a menu page, simply highlight the function using the cursor key, and press the ENTER button.

**Cursor Key.** This key is usually a four-position thumb key that is used to scroll around the various pages of the receiver and highlight the functions you wish to access. It performs the same basic function as the "mouse" of a personal computer. It is also used to enter data in selected fields on various pages of the receiver. For example, you use the cursor key to name a waypoint by pressing the key left or right to highlight a position where you wish to start the name, pressing ENTER to indicate you wish to input the name, and then pressing the cursor key up or down to select the specific character you wish to use. We will discuss this procedure in more detail in the waypoint management section.

**Page Button.** This button allows you to scroll through the various main data pages of the receiver. It will also return the display from a sub-menu page to the previous page viewed. On some GPS receivers, this button is also used to display messages when a message alert occurs.

***Mob Button***. MOB stands for Man Overboard. Pressing this button causes your present position to be recorded in the receiver's memory, and steering directions are immediately displayed to allow you to return to the exact location where the button was activated. Some receivers do not have a separate MOB button, but use either the **Mark** or **Nav** button as a secondary feature to activate the MOB. In this case, the MOB function is activated by either pressing the **Mark** or **Nav** button twice, or by holding the button down. Your user manual can give you specific information on the use of this feature.

***Nav Button***. Not all receivers have this feature as a separate button. Normally, pressing this button will bring up the Navigation Page. On some GPS receivers, this button is used instead of the GOTO button.

***Quit Button***. This button is used to return to a previous page or to clear a data entry field. It is analogous to the "Escape" (Esc) key on a personal computer. Some receivers label this button as **Clear** rather than **Quit**.

The buttons described here represent a generic GPS receiver. Your receiver may have more or less than those described. Figures 3.2 and 3.3 show the layouts for two basic handheld GPS receivers.

Again, you are encouraged to review your user's manual for specific information about the buttons on your receiver. Now that we have looked at some of the buttons available and their use with the GPS receiver, we need to turn our attention to the various pages and menus available and how we can use them for navigation. But, before we leave the buttons, let's state one more time, don't be afraid to press the buttons and explore what they do. You won't "break" your receiver, and you just might get a better perspective of how your receiver works, and what these buttons can do. You are

*Figure 3.2 – Layout of one basic handheld GPS receiver model.*

*Figure 3.3 – Layout of another basic handheld GPS receiver model.*

encouraged to experiment with your user's manual in hand. Hands-on is one of the best ways to learn how to use your GPS receiver.

## PAGES

We have already taken a look at the first page that our receiver will display when we turn it on, and the information that we can obtain from just that one page. Now, let's look at the other many pages that are available when using our GPS receiver for navigation. Let's first look at a page that will give us the information we need in order to determine our exact present location.

*Position Page.* Figure 3.4 shows a typical position page on a handheld GPS receiver. Note at the top of the page there is what looks like an analog ribbon compass, but it is not a compass. This is confusing to many people because every GPS receiver has a page that looks like a compass display. The GPS receiver can only tell you the direction in which you are moving, not the direction in which the receiver is pointed. The diamond in the center of the display indicates your present course over ground, not the direction in which you are pointing the receiver. In GPS language this is your "track". From the setup menu you can set the direction indicated here to either true or magnetic north. We will discuss the setup menu later. The Position Page also indicates your present latitude and longitude as well as GPS time. The time can be adjusted to either local time or Coordinated Universal Time (UTC) in the setup menu. Speed is also usually indicated on this page. GPS determines both the direction and speed in which you are moving by comparing your present position with your earlier position and computing the course and speed from the change in position over the elapsed time.

Other data fields are also presented on this page. This particular model GPS receiver in Figure 3.4 has two user selected data fields, trip odometer and altitude. Some other user selected data

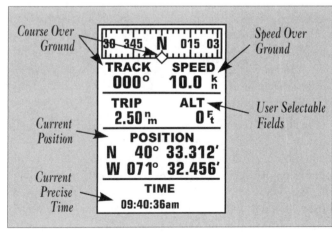

*Figure 3.4 – Position Page.*

fields include elapsed time, average speed, maximum speed, and trip timer. Not all GPS receivers offer a Position Page *per se* as shown here, but the information is available on one or more of the other pages. For example, the Garmin GPS 76 displays the position information on the Satellite Page. See Figure 3.5.

*Map Page.* On the map page your current location is indicated by a symbol such as a cross or a triangle (some units let you select the symbol). A dotted line trail indicates your past course over ground if the track function has been activated. Waypoints and other objects whose coordinates have been

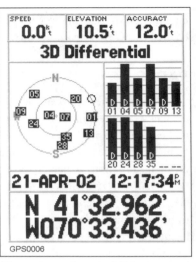

*Figure 3.5 – Garmin GPS 76 Position Page.*

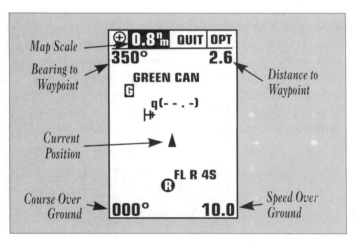

*Figure 3.6 – Typical Map Page.*

preloaded into the GPS receiver also will appear if they are within the field of view of the current screen. The field of view usually is selectable with a zoom feature. Zoom is controlled either with dedicated buttons or through menu options on the Map Page. Figure 3.6 represents a typical Map Page in which the data fields indicate the course over ground and the speed over ground. Two other data fields indicate the bearing and distance to a waypoint that has been selected. The display also indicates the level of zoom and scale in miles of the map display. The Map Page can be one of the most useful displays on a GPS

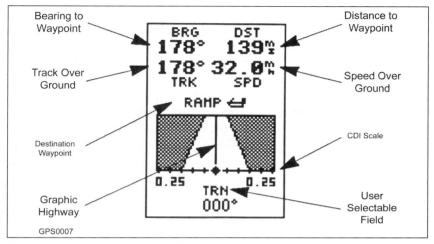

Figure 3.7 – Highway Page of a GPS receiver.

receiver, especially if utilized effectively. To be useful, this page needs to have objects other than your present position displayed. The Map Page will show all waypoints entered into the GPS receiver that are within the field of view of the current screen.

*Highway Page.* Figures 3.7 and 3.8 show the highway pages of two different GPS models. The Highway Page shows a 3-D depiction of your position on an imaginary highway. The viewpoint is from a spot somewhat above your current location and looking in the direction you are moving. The centerline on the highway represents the original course line drawn when you selected the active waypoint. In Figures 3.7 and 3.8 you are shown to be "on

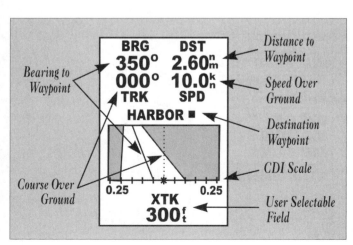

Figure 3.8 – Highway Page of a different GPS receiver.

course". The centerline of the highway lines up with the center of the display, and it appears as if you're located in the middle of the highway. The highway stretches straight ahead of you indicating that you are heading directly toward the selected waypoint. If you were to the right of the intended track, the center of the highway would appear to be off to the left as shown in Figure 3.8. The scale below the highway shown on this GPS model is the course deviation indicator (CDI), or the distance to which you are either left or right of the highway centerline. The scale is user selectable. Many models have eliminated this data field because the "crosstrack error" data field shown in Figure 3.8 and represented by "XTK" provides the same information. The Highway Page also provides the bearing and distance to the selected waypoint in addition to the current track (course over ground) and speed over ground.

## MENUS

The menu function of the GPS receiver provides the primary means of controlling the receiver. Each receiver has multiple sets of menus and submenus that are used to select the functions you wish the receiver to perform. One set is used to access the navigation functions of the receiver, such as waypoints and routes. A second set is used to control how and what the GPS receiver computes and displays. A third set controls the interface between the GPS receiver and other devices you may wish to use with your receiver, such as downloading/uploading waypoints from a personal computer. Perhaps the menu function offers the greatest source of confusion and frustration to the first-time user as different GPS models have a different way of accessing the menu function. The starting point is something called the Main Menu Page.

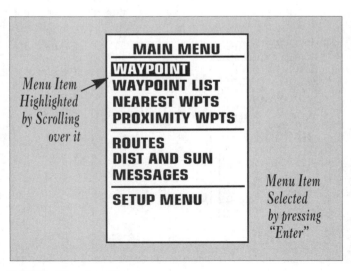

*Figure 3.9 – Typical Main Menu Page.*

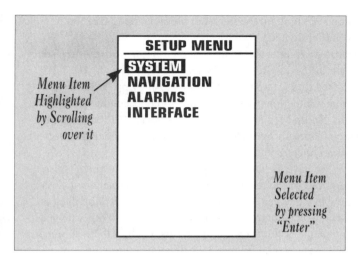

Figure 3.10 –
Typical Set-
Up Menu.

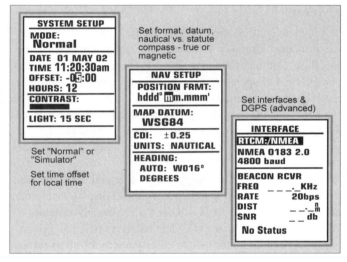

Figure 3.11 –
Set-Up Menus
of several
different
GPS receiver
models.

Some receivers have a button labeled **Menu** which allows direct access to the Main Menu Page. Others require you to access the Main Menu using the **Page** button where you must page through the various screens until you reach the Main Menu Page. Figure 3.9 shows a typical Main Menu Page. The Main Menu Page is where you access most of the other functions which are basically:

   **Waypoints** - the destination point used for navigation;

   **Routes** - sequences of waypoints for navigation;

   **Setup** - controls the receiver and is usually divided into submenus labeled System Setup, Navigation Setup, Interface Setup, and Alarms Setup. Items are selected by scrolling to highlight the item and pressing the ENTER button once the selected item is highlighted.

We will discuss waypoints and routes in detail in the next chapter, so let's concentrate here on the Setup Submenu. When you first turn on your GPS receiver you will need to set it up to function and display properly depending upon your desires and intentions. To do this we would select the Main Menu Page and highlight "Setup Menu". When we press the ENTER button on the keyboard a submenu labeled "Setup Menu" will appear on the display screen. Figure 3.10 shows a typical Setup Menu.

Note that now we have a choice of a system setup menu which is used to control the operations of the receiver, a navigation setup menu which is used to control how the receiver computes and displays navigation data, an interface setup menu which controls how the receiver talks to other devices, and an alarm setup menu to control warning alarms. Each of these submenus is accessed through the setup menu. Figure 3.11 illustrates the setup submenus of several GPS receiver models.

## System Setup Menu

The system setup menu controls the operation of the GPS receiver. It first offers a choice of mode: Normal or Simulator. Normal mode is the active mode for GPS navigation. Simulator mode turns off the receiver's antenna and active navigation operation and operates the receiver with simulated satellite signals, usually captured at the last time the receiver was in Normal mode. The simulator mode can be invaluable in planning and practicing with the receiver. When using the receiver for actual navigation, however, it is important to ensure it is in Normal mode. To make sure this is the case, most GPS receivers return to Normal mode each time the set is turned off.

The System Setup Menu allows the user to set the date and the offset from Coordinated Universal Time (UTC) so that the receiver will display local time. The GPS satellites operate in UTC and this is what will be displayed unless an offset is set in the receiver from this menu. Most receivers will also permit the user to select either a 12 or 24 hour clock to be displayed as well as set the contrast level on the display. As on all the other pages, you can select these fields by moving the cursor up or down using the cursor key until the desired field is highlighted, then pressing the ENTER button.

## NAVIGATION SETUP MENU

The Navigation Setup Menu allows the selection of format for displaying position information. Most navigators select degrees, minutes, and tenths of minutes because the latitude and longitude scales on charts and maps use these same formats. Using these measurements allows direct transfer of location from the receiver to electronic charts and maps as well as facilitating easy plotting on paper charts and maps. In this menu you will need to set up the map datum you use for navigation. Let me explain what is meant by map datum. The Earth is not a perfect sphere, in fact it bulges more at the equator than at the poles. This is a shape known as an oblate spheroid. In order to draw the circles of position (COP) on the surface of the Earth based on the distance from each satellite, the GPS receiver must have a reasonably accurate model of the Earth's shape. This model is known as a datum. Over the years geodetic scientists have measured and refined this model. The latest version is known as World Geodetic Survey 1984 (WGS-84). Most modern day charts and maps use this model. It is essential that the datum used by the GPS receiver to calculate positioning information correspond to the datum used by the electronic or paper map or chart you have selected. For nautical charts, the datum is listed in the title block of each chart. Most modern day road maps use WGS-84 datum. There are a number of other datum grid systems used by foreign countries and other applications such as the US Geological Survey charts. Most GPS receivers can calculate positioning information using these systems, but you must select the system from a list carried in the receiver's menu. The default datum setting for most GPS receivers will be WGS-84.

In the Navigation Setup Menu you can select the measurement units in either statute miles or nautical miles depending upon whether you are navigating upon the land or water. Most receivers will also permit you to select from this menu whether you wish your heading information in true or magnetic direction. If you are using a compass to assist in navigating you may wish this information in magnetic direction to correspond to your compass reading. If you select magnetic direction the GPS receiver will automatically insert the proper degree of variation for your stated position.

## INTERFACE SETUP MENU

The Interface Setup Menu establishes the protocol for data exchange. This allows you to connect your GPS receiver to other devices for the exchange of data. The most commonly used protocol is National Marine Electronics Association 0183 (NMEA 0183). We will discuss interface protocols and the connection to other devices in Chapter 7.

## ALARMS SETUP MENU

The Alarms Setup Menu enables the user to turn various alarms on and off. Typical alarms are waypoint arrival alarm, off course alarm, and proximity to waypoint alarm. Most GPS receiver alarms have selectable thresholds for distance, and the values are entered using the cursor key.

# 4 | Navigating with the GPS Receiver

Now that we have a familiarity with how our GPS receiver works, we need to learn how to use it to navigate. Using the mark button on our receiver we can establish a position on all the places we have been, but if we wish to do more than just return to those places we have been, we need to understand how to enter waypoints in our receiver. In order to do that, we need to know something about latitude and longitude, the language of navigation.

The system of navigation called *latitude and longitude* was developed by the early sailors to help guide them as they sailed into the unknown waters of the Earth's oceans. This system gives every individual spot on Earth a unique address described as coordinates of north or south latitude and east or west longitude. Latitude and longitude is simply a way of describing a unique position anywhere on Earth. If you could imagine the Earth divided with imaginary grid lines that flow north and south of the Equator as lines of latitude, and east and west of Greenwich, England as lines of longitude, it would look something like Figure 4.1. In this imaginary grid system the Earth is divided into lines of latitude and longitude that determine how far north or south of the Equator you might be and how far east or west of something called the *Prime Meridian* you are.

To give each location

*Figure 4.1 – Earth with grid lines overlaid.*

on Earth an address a numbering system has been defined using degrees and minutes of degrees. In the case of latitude, the Equator in the center of the Earth is considered at 0° latitude and the North Pole at the top of the Earth as 90° north latitude. Likewise, the South Pole at the bottom of the Earth is considered at 90° south latitude. When determining latitude the respective hemisphere must always be indicated as either north or south, otherwise you wouldn't know if you were describing a location in the United States or one in Argentina. All latitudes are parallel to the Equator; consequently, they are called *Parallels of Latitude* or just simply Parallels.

To describe longitude a starting location had to be established. The English were the dominant navigators of the 18th century and choose to define the starting point as the Royal Observatory of Greenwich, England. This definition was debated by the various nations of the world until finally it was officially adopted by international agreement at the 1884 International Meridian Conference in Washington, DC, and Greenwich, England became the Prime Meridian, the starting point for all lines of longitude. Lines of longitude are called *Meridians* and are great circles that run from pole to pole east and west of the Prime Meridian.

Longitude is a number that runs from 0° at the Prime Meridian east and west to 180° at the International Date Line. Like latitude it is necessary to specify the hemisphere either east or west when describing longitude. So that there is no confusion, note that all locations in the United States are North Latitude and West Longitude.

To give you an idea of what a location would look like when addressed in latitude and longitude, check out the location of ARRL Headquarters. On your GPS receiver the coordinates would appear as N 41°42.882′ and W 72° 43.650′. That's all well and good, but unless you had a chart or a map with latitude and longitude grid lines you would never know where in the world this place was located. This brings us to our next discussion about charts and maps.

As noted in Chapter 3, the Earth is not a perfectly round sphere. It is more of an ellipsoid with a bulge in the middle at the Equator and a flattening at the Poles. This presents a problem when trying to determine a location on the Earth using latitude and longitude because this system is designed to work with a perfect sphere. To compensate for the ellipsoid shape of the Earth and the latitude/longitude system, a model was devised which compensates for the differences. This model is called a *datum*. Over the years many models or datums were devised and the latest and most accurate is the one based on the world geodetic survey accomplished in 1984. This datum is known as *WGS 84*. Prior to WGS 84 a geodetic survey was done for North America in 1927 and known as *NAD 27*, and many topographic maps still carry this datum. Almost all GPS receivers carry the default datum of WGS 84. It is always best to check the chart or map you are using to determine which datum was used in its manufacture. If you find that datum is something other than WGS 84, it is easy to change the datum in your receiver by going to the navigation set-up menu and selecting map datum or datum. Most GPS receivers will list a wide variety of datums that can be selected, but unless you are an archeologist or geodetic scientist using some ancient map or chart, you will only have need for WGS 84 or NAD 27 as your map datum.

Now that we know all that we need to know about latitude and longitude and map datums, we need to turn our attention to how we use that information with our GPS receivers. One of the most basic concepts of GPS is something called *waypoints*. Waypoints are simply an address of where we want to go. We have a starting address or waypoint and a final destination waypoint as well as intermediate waypoints along the way between our starting point and our final destination. Once we have all the waypoints put together between our starting point and our final destination we can bundle them all together into something called a *route*. As noted earlier, the GPS informs us of our position in the language of navigation, which is latitude and longitude. To tell the GPS receiver where we want to go we have to tell it in the language it understands, latitude and longitude. A modern GPS receiver can store upwards of 500 waypoints in its memory. We enter waypoints, the place we want to go, into our GPS in the coordinates of latitude and longitude. There are several ways we can do this and the simplest is to go the location where we wish to mark as a final destination and push the buttons on our receiver to tell it to store that location in its memory. But what if we wish to go to a location where we have never been before? That's a little more complicated and requires that we already have a set of coordinates for that location. The coordinates might have been given to you by a friend, or you found them in a travel book, or you took them from a map or a chart. In any case, the GPS will require that you tell it where you wish to go in latitude and longitude. There are several ways we can do this.

## MANUAL ENTRY

For older GPS receivers that do not have a mapping capability or for basic units also without a mapping capability, this is the only way to input coordinates into your GPS outside of using the mark function. This method requires inputting characters and numbers into the GPS receiver one at a time. In order to enter waypoints manually you must access the Waypoint Screen. On most GPS Receivers you can do this by going to the Main Menu and selecting "waypoints" or "waypoints list". Using the cursor key scroll over and highlight this field and press ENTER. A new screen offering the option for

## WAYPOINT LIST

**NAME** — — — — — — —

— — — — — — —

— — — — — — —

**REF:** — — — — — —

**BRG**                    **DST**

**RENAME ?**              **NEW ?**

**DELETE ?**              **DONE ?**

*Figure 4.2 – Typical waypoint screen.*

new waypoints will appear. Figure 4.2 is a typical new waypoint screen. Notice there is a place for a name and below the name a place for the latitude and longitude of the new waypoint. Figure 4.3 is a typical new waypoint screen filled in with the required information. Once on a new waypoint screen, here's how you would enter the information: First enter the name you wish to give the waypoint. Using the cursor key, scroll down and highlight the name field. Press ENTER to select it. Using the cursor key press right to select the character

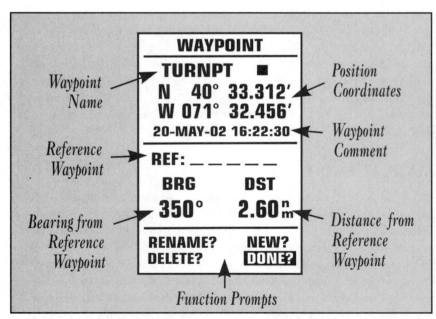

*Figure 4.3 – Typical new waypoint screen filled in with data.*

field, then scroll up or down to select the first character. Continue until you have entered the name you wish to give this waypoint. Press ENTER when finished.

Next we have to enter the waypoint coordinates. On the same screen scroll down to the navigation field which should be directly below the name field. When this field is highlighted, press ENTER to select it. Now using the cursor key press right to select the character field and up or down to select the character. Continue until you have entered both the latitude and the longitude of your waypoint. Don't forget to add either "N" (North) or "S" (South) for the latitude and "E" (East) or "W" (West) for the Longitude. When you have everything entered on the screen the way you want it, scroll down to the bottom of the page and select "done" and press ENTER. Using the manual entry method this is the way we must enter each and every waypoint into the GPS. As you can see, this method is rather labor intensive and prone to much error. There is a much easier way if you have a mapping GPS receiver, though.

## SCROLL ENTRY

This method requires that your GPS receiver have a screen that can display a map or chart picture of the location where you wish to go. This is called a *mapping GPS*. On most mapping GPS receivers using the manual entry method is very difficult or impossible and the scroll method is preferred. The map or chart used in mapping GPS receivers is an electronic version of a paper map or chart that is separate from the receiver and usually contained on a card or cartridge that is inserted into the receiver. Some typical handheld mapping GPS receivers are the Garmin GPSMap 60CE, the Garmin GPSMap 76 the Lowrance iFINDER H20c, and the Magellan Meridian. Some mapping GPS receivers can also receive their electronic map or chart data by downloading it from a personal computer.

Regardless or how the maps or charts are loaded into the GPS receiver, in order to use the scroll method the location you wish to go to must be displayed on the GPS screen. Establishing a waypoint then becomes a simple matter of using the cursor key to scroll to the area where you wish to establish a waypoint and pressing the "mark" button. A waypoint screen will be instantly displayed with the coordinates of latitude and longitude of your waypoint and a numerical name that the GPS has given it. If you wish to change the name you can do so at this time or you can choose to do so at a later time. The GPS will sequentially number the waypoints as you establish them starting with the lowest number available. You next only need to accept the waypoint by selecting "Save" or "Done" at the bottom of the waypoint screen and the GPS will store the waypoint within its memory. You can continue in this fashion scrolling to areas of the electronic map or chart selecting waypoints until you have your whole route laid out.

Some mapping GPS receivers that are designed for use in automobiles have a built-in map of the principal roads, major city streets and Interstate Highways or they come with computer software of electronic maps that is downloaded to the GPS using a cable connection between the GPS unit and the computer. These mapping GPS offer what is called *real-time routing*. Basically this means that you enter a destination directly into the GPS receiver and it calculates your route. These receivers can vary, however, in terms of how you enter a destination and what kinds of destinations you can choose. Some require you to enter an address, usually the city, street, and street number. Some in addition to the address destination will allow you to enter a point of interest or an intersection. Most will allow you to scroll over a map from your current position until you find where you want to go then mark that position on the map and use the software's "Route To" function. Bear in mind these special types of GPS receivers only will route you to city streets, roads, highways, and Interstate Highways. If you wish to use your GPS to hike outdoors, camp, hunt, fish, navigate on the water, find the perfect field day location, or participate in geocaching a specialized GPS designed for use in automobiles will be of little value.

## MARK ENTRY

Marking waypoints is by far the easiest method for entering waypoints. While you are underway and passing a point of interest that you would like to include as a waypoint, simply press the "Mark" button on you GPS. This will bring up a waypoint screen which will have a number assigned by the GPS for a name and the exact coordinates of the location in latitude and longitude. You need only to accept or change the name, like in the scroll method, and press "Save" or "Done" to include this waypoint in your GPS unit.

## COMPUTER ENTRY

Most modern GPS units have a data port that can be connected either to the serial port or a USB port of a computer using a cable supplied by the GPS manufacturer. These cables range in price from $20 to $50 depending upon the model GPS receiver, and can be purchased from dealers, on the World Wide Web, or directly from the manufacturers. The other component needed is the software for the computer. This also is readily available from the GPS manufacturers as well as some software vendors such as MapTech and Delorme. Using the computer you can not only upload waypoints to your receiver but also download stored waypoints from your receiver to the computer. Using the computer to off-line plan a trip or an activity is highly recommended as it eliminates errors that might be introduced with other entry methods and allows you to easily make edits to your route plan. Since each software supplier has unique requirements for entering waypoints, to use this entry method you will need to read and follow the instructions that come with the software.

Once we have the coordinates we think we want entered in the GPS, we may want to go back and check them to be sure they are exactly what we want. One of the nice features of the GPS is we can edit waypoints and routes at will. When you wish to edit a waypoint simply bring up the waypoint list from the Main Menu screen, highlight the waypoint you wish to edit and press ENTER. This will bring up the waypoint screen of that particular waypoint. This is the same waypoint screen you used to enter the waypoint data. Using this screen you can now edit the name, the coordinates, and the reference data that you entered. By scrolling over the "Save" or "Done" icon and pressing ENTER the updated waypoint information will be stored in your GPS receiver.

If you used a computer to plan your route and uploaded the data from the computer to your GPS, you the have perfect platform to perform editing functions. On the computer bring up the route you wish to edit and using the computer mouse click on any waypoint you wish to change or adjust. You can delete waypoints, change their name, move them on the electronic map or chart, or add waypoints all with the press of a few keys and the click of the mouse button. Check with the instruction manual that came with the software to get a good insight of the capabilities of your particular software for computer editing.

Entering waypoints in the receiver is the first step in planning where you want to go. The next step is to arrange the waypoints into a route. A route can be a trip that you want to take, or paths to areas you frequent, or a search for a special field day spot, or any place you wish to go. The key feature of any route is an analysis of the best waypoints to select to safely take you from your present location to where you want to go. Bear in mind that the GPS will always guide you from waypoint to waypoint in a straight course. You must make sure that the course you have chosen is free from obstacles and hazards because the GPS has no knowledge of your environment or surroundings. You should make sure that a waypoint is entered for each and every point along your planned route where an action should be taken, such as turns, intersections, or any point where a decision must be made. As you enter the waypoints to make up your route the GPS will compute the course segments and distances for each leg of your route. If you are using a paper map or chart to assist in your route planning you might want to annotate it with this information.

To begin establishing a route, access the Route Menu from the Main Menu by scrolling over and highlighting "Routes" and then pressing ENTER. This will bring up the Route Definition screen. See Figure 4.4. The GPS receiver will display "Route 1", if this is your first route, and a name for the new route. Below the name and route number is a set of blank fields for listing waypoints and other information. Scroll over the first empty field and press ENTER. This will activate the first waypoint field where you wish to enter the first waypoint to start your route. By scrolling first the left character field then successively the other fields to the right you will bring up the waypoint names you stored in memory. Continue in this fashion until you have brought up the desired waypoint you wish to start your route. To accept this waypoint as the one you want, press ENTER. The Route Definition screen will display the name of the first waypoint. The distance will appear as 00.0 since is the starting point of your route.

Next, scroll over the waypoint field immediately below the first one and repeat the process. You will enter the second waypoint in the route in the same manner as you did the first one. Once you have the second waypoint established, a course leg will be established and the distance and course fields will have values filled in to reflect the distance and course heading from the first waypoint to the second. Repeat this process until the entire route is filled

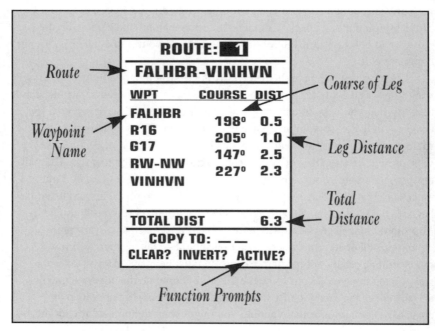

*Figure 4.4 – Typical route definition screen.*

in. The distance field will show the distances between each successive leg of the route and a total distance for the route at the end. The route can now be activated and used and the route name will now appear as some combination of the first and last waypoint names. Most modern GPS receivers will allow you to change the name by highlighting it and pressing ENTER, then enter the name the same way you would enter a name for a waypoint. The only difference is waypoint names are limited to 6 or 8 characters, whereas route names can have up to 15 characters or more.

Once waypoints are entered in your GPS receiver it is easy to access them to navigate to your desired location. There are several ways to do this.

• *Use a route.* Using a route is the easiest way to access waypoints. Once a route is established within the GPS receiver it can be activated by entering the Main Menu and selecting "Route". You can then select the route you wish from the list you previously established by highlighting the route name and pressing ENTER. This will bring up the Route Definition Page where you can select from four options. "Act" or "Activate" to start the route at the first waypoint, "Inv" or "Invert" to start the route in the reverse order of waypoints, "Clear" to remove the route from the list, or "Copy To" to copy this route or selected waypoints from this route to another route. The invert feature is a handy to use when you are traveling in unfamiliar territory. At any point along the route you can reverse the waypoints by selecting the "Inv" option and follow them back to your original starting point. This is commonly called the "*bread crumb*" feature after the famous story of Hansel and Gretel. Another great feature is the "Copy To" command. You can use this command to copy commonly used route segments into longer routes and save the effort of repeatedly entering the same data. You can also edit a route to add or delete waypoints with this feature.

• *Use the "Go To" button.* By pressing the Go To button on your receiver the waypoint list of all the waypoints you have entered into your GPS will appear. Simply select the desired waypoint from the list by highlighting it with the cursor key and press ENTER. The receiver will set a course from your current location to that waypoint. The receiver will draw a line on the Map Page screen from your current location to that waypoint and also create a corresponding highway on the Highway Page screen. Some GPS receivers have a "Nav" button instead of the "Go to" button. If your receiver has a Nav button you need to press that key to get a sub menu that usually contains "Go To" and other navigation functions such as "Find" and "Route". To get to the selected waypoint follow the course laid out by the GPS. You can continue in this fashion by pressing the Go To button as you reach each successive waypoint or you can group the desired waypoints into a route as explained earlier and the GPS will automatically set a course to the next waypoint in succession until you reach your final destination.

• *Scroll over a waypoint.* To use this function, scroll over the map page to the desired location and press "Go To" then press ENTER. The GPS will set a course from your present location to the location you selected. This is a handy feature when you want to change your course for some reason and pick up a new heading or change to a new route.

Once a route is activated, the GPS will start navigating to the first waypoint and display information of the Map Screen, the Highway Screen and the Position Screen. From this point on you only need to follow the course directions displayed on these screens. You may find that the Highway Screen is one of the most useful for navigation purposes as all one has to do to stay on course is to keep centered in the middle of the highway displayed. If you go off course, the GPS will alert you to this fact and tell you how far off course you are, as well as point you to the direction you must steer to return to your original course. Refer to Fig. 3.6 in Chapter 3.

Basic navigation with GPS becomes a matter of selecting and entering waypoints in the receiver, grouping the selected waypoints into a route, and activating the route. Follow the route and you are navigating.

# 5 | Using GPS with Topographic Maps

In GPS navigation maps and charts are a very useful and necessary part of the equation. With the right map or chart you can find the latitude and longitude of places you may want to visit, plot a course across the land or on the water, identify obstacles along your path, and verify your position on land or water by comparing the actual physical features of your surroundings to their representation on the map or chart. In this chapter we will discuss those maps we can use on land that can detail the natural and manmade features of the surroundings we choose to visit.

We are all familiar with the highway maps that we use to plan our automobile trips. These are called *planimetric* maps. They show highways, towns, and cities but they don't tell us anything about the terrain. Looking at these maps you can't tell if the land is flat or a steep mountain grade. In addition, they don't list the latitude and longitude of any location, something you will need to program your GPS. Highway maps are great for highway use in our automobiles, but if we want to hike, camp, or venture into the great outdoors we need something called a *topographic* or "topo" map.

## CONTOUR LINES

A topo map not only shows details that are not shown on a planimetric map, it also shows the shape of the terrain or the surface characteristics of the Earth. It does this by the use of contour lines which are imaginary lines that join points of equal elevation on the surface of the land above and below a reference point such as mean sea level. Contour lines on a topo chart make it possible to measure the height of a mountain, the depth of a valley, the steepness of a slope, or even the depth of the ocean. Contour lines are printed on a topo map in brown ink and every fifth line is printed in bold and marked with its elevation. This fifth contour line is called the *contour index*. The shape of the contour lines also reveals a great deal about the terrain they represent. Widely spaced contour lines represent gentle slopes, whereas closely spaced lines represent steep slopes. Mountain peaks and land areas that come to a point are represented by concentric rings, V or U shaped lines pointing downhill represent a ridge, and V shape lines pointing up hill represent a canyon or gully.

Now take a look at Figure 5.1. To illustrate how contour lines reveal the shape of the Earth's surface we will use a portion of an actual topo map. Note the words "Contour Interval 20 Feet" at the top of the page. Actually this notation appears on the lower center margin of the map but it is placed here to give you a reference to the elevation change between contour lines. The line at "D", for example, is 20 feet below the line at "A". Obviously, a place on the map with a lot of contour lines has a large amount of elevation change. Take a look at the area around "C". This is a steep hillside with a large amount of elevation change. Compare that to the area around "B". This is a flat area with no elevation change. Areas with no or very few contour lines have little elevation change and are flat. Note that every fifth contour line is bolded and if you follow it you will eventually find a place where the elevation is given. These are called Index Contours. The contour line at "E" is such a place. Sometime topo maps will contain "spot elevations". These are places on the map that are marked with a small "x" and the elevation is given for that spot The location at "F" is a spot elevation. Contour lines that end in a circle are peaks and are sometimes marked as spot elevations. The small circle and "x" at the bottom right of the map marked 7901T is such a location. Disregard the "T" and the elevation is 7901 feet.

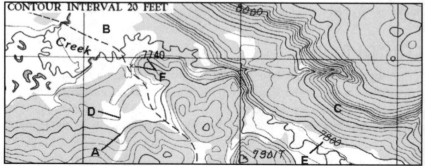

Figure 5.1 – A portion of an actual topo map used to illustrate the contour line features and to explain how they relate to terrain elevation. Courtesy of United States Geological Survey.

# TOPO MAP SYMBOLS

In addition to the contour lines a topo map includes symbols that represent features such as buildings, streams, streets, roads, trails and vegetation. Area features appear in color on the map. For example, vegetation is always green, water is blue, and densely built-up areas are either red or gray. Individual houses may be shown as black squares; whereas large buildings may be depicted in their exact shape. These symbols are constantly being refined by the United States Geological Survey (USGA) to improve the appearance and readability of the topo maps. Unfortunately, this means that within the same series of maps there may be slightly different symbols for the same feature. The amount of various symbols used on topo maps is too numerous to include here, but to help with the interpretation of these symbols the USGS publishes a booklet entitled *Topographic Map Symbols* that is free for the asking at 1-888-ASK USGS or you can download a copy at **erg.usgs.gov/isb/pubs/booklets/symbols.**

# ELEVATION

As mentioned in the previous paragraph, elevation is referenced to mean sea level. That's good as long as you are near the sea. But, what do you do if you are in the Rocky Mountains or in the deserts of Utah or Arizona? Topo maps define mean sea level, when you don't have the sea handy to reference as the elevation at which the force of gravity is the same as at actual sea level. This is the zero reference level or vertical datum. It is defined as a line of constant gravity or geoid. The *geoid* is not a level surface but it varies with the shape of the land and the density of the rocks in the immediate vicinity. In mountainous areas the mass of the mountains makes the referenced mean sea level higher than it would be if the mountains were not there. In desert areas referenced mean sea level can be below the surface level of the surrounding terrain.

Another way to determine mean sea level is to make it into a mathematical model representing the earth as an ellipsoid. Mean sea level then becomes the height of the ellipsoid that best represents the true shape of the earth. The model which is the default model for most GPS receivers is the WGS 84 *vertical* datum. GPS receivers work with both a horizontal datum and a vertical datum, and in both cases the usual default datum is the WGS 84. Mathematical models work well for the GPS receiver but they only approximate the true geoid. Across the United States the true geoid ranges from about 10 feet above to 100 feet below the WGS 84 vertical datum. This explains why the vertical accuracy of the GPS receiver is no where near the horizontal accuracy. The most recent geoid of the continental United States based on millions of gravimetric measurements is GEOID99. Modern topo maps use the GEOID99 as the reference for elevation.

# READING TOPO MAPS

Although topo maps are produced by many private companies, the most detailed and useful are those produced by the US Geological Survey. USGS maps are available online or by mail order directly from USGS, and from many outdoor shops that carry hiking and camping equipment. To order directly from USGS online go to their Web site at **www.topomaps.usgs.gov** and search for the particular area you are interested in. The USGS creates several scales of topo maps, but the most detailed scale is the 7.5 minute map scale (1:24,000). 7.5 minute maps are so called because each covers 7.5 minutes of latitude and 7.5 minutes of longitude of the Earth's surface. This is approximately eight miles north and south by six miles east and west. Each 7.5 minute scale paper map is about 28 inches long by 21 inches wide. There is a wealth of useful information available on the borders or margins of every topo map that can help us in understanding the map. Starting with the right lower border we find the name of the map, its publication date, and a small legend that can help us read road and highway symbols. In the center lower border you will find the contour interval, the map scale and the vertical datum. Coastal area maps also include the water depth and shoreline data. Note: the datum referenced here is the vertical datum not the horizontal datum. Don't confuse the two. On the lower left border there is a lot of information, mostly hydrographic data, but it is here you will find the horizontal datum that you must ensure matches the datum in your GPS receiver. On most topo maps this will be North American Datum 27 or NAD 27. On these maps that were surveyed to NAD 27 there will be instructions on how to translate coordinates to the WGS 84 datum, but it may be easier to just reset your GPS receiver to NAD 27.

Around the borders of the topo map are something called *neatlines*. These are the lines on the map that align with either an even minute or a 30 second line of latitude or longitude. The latitude neatlines run up the sides of the map, whereas the longitude lines run along the top and bottom of the map. Both latitude and longitude are shown in degrees, minutes, and seconds. The full numbers with degrees are shown in the corners of the map but only the minutes and seconds are shown in between the corners. Unfortunately, numbers for other coordinate systems are displayed between the neatlines sometimes making it difficult to pick out the minutes and seconds of latitude and longitude. You can read the latitude and longitude of any point on the map and enter this number in your GPS as a waypoint. Or, if you have a mapping GPS receiver with a topo map displayed, you can scroll over a point on the map and enter it automatically as a waypoint using the techniques described in Chapter 4. Of course, this presupposes that you have invested in a mapping GPS receiver, have obtained software topo maps for the areas you are interested in, and have downloaded or installed them in the GPS receiver. Since more and more of the GPS receivers offered on the market today at a reasonable price have a mapping capability, this has become the popular choice of those who want to venture into the great outdoors, both for planning purposes and for real time navigating. We will look at GPS selection in Chapter 7, but for now let's look at what is available in software topo maps.

## SOFTWARE TOPO MAPS

Paper topo maps are excellent for use in exploring the outdoors, and if we have only a non-mapping GPS to assist us they are the best available. There are some drawbacks to paper maps, though. For one thing they don't cover a very large area and if your exploring is going to cover any distance you will have to carry several maps. For another, extracting latitude and longitude information from a paper map and plugging that information into your GPS is tedious work and prone to error. A better way is to use electronic or software maps. The area covering a whole state can be contained on one software CD-ROM or custom chip, and it is possible to move from one area to another seamlessly within that State without going to another map. Establishing waypoints can be as easily as moving a cursor key to position a pointer on the GPS screen to where we want the waypoint to be and pushing a button.

There are many types of software maps and multiple sources for each type. There are those that are designed for streets and highways that are the planimetric type and are mostly used in automobiles, and there are those called charts that are designed strictly for use by mariners on the water. Then, there are those that are designed for use in the outdoors on land and are called topographic maps. We will limit our discussion in this Chapter to these types of maps, the software topo map.

Some software topo maps like the Garmin *MapSource* or the Magellan *Mapsend* are designed to load directly into a compatible mapping GPS receiver. These types of programs allow you to use the map software directly in real-time navigation as well as use it in planning a journey on a personal computer. Both companies, however, use a proprietary protocol that allows only the use of their software on their GPS receivers. You can visit Garmin online at **www.garmin.com** and Magellan at **www.magellangps.com**.

Other programs like the National Geographic *TOPO!* and *Terrain Navigator* from Maptech are designed to run only on a personal computer. Both of these programs allow you to plan your travel in detail directly on the PC screen and then transfer your waypoints from the PC to a GPS by the use of a serial or USB connection. The only thing you cannot do with these programs is download the maps to the GPS. The maps can be downloaded only to Microsoft *Windows* compatible software device such as a laptop computer or handheld pocket PC. Both of these programs are complete turnkey systems with both maps and software interface to operate the computer. Simply load the program into your computer and you are ready to use all its features. You can investigate Maptech at **www.maptech.com** and National Geographic at **maps.nationalgeographic.com/topo**.

# 6 | Ham Radio Applications with GPS

The GPS receiver is an excellent positioning and location tool that has found its way into many facets of our life today. We find its application in everything from automobiles to airplanes. It is even turning up in our cellular telephones, cameras, and wrist watches. It won't surprise you to learn that it also has many applications in ham radio. Some of these ham applications are just plain fun while some serve a serious and most useful purpose. Let's examine some of the ways GPS and ham radio are congenial partners.

## GEOCACHING

Let's start with a fun activity where hams can use their handheld VHF or UHF radios and a GPS receiver in a game that is a variation of *geocaching*. But first, what is geocaching? It's been called everything from a scavenger hunt to a high-tech game of hide and seek. The term comes from the joining of two familiar words: geo, which refers to the earth, and *caching*, which comes from the word cache referring to a hiding place which is used to temporary store items. The concept of geocaching is relatively simple. One person hides a container that typically includes a log book, an instruction sheet, and an object or trinket of some kind. The coordinates of this "hidden treasure" or cache are then posted on an Internet Web site. Others, using their GPS receivers and the posted coordinates, try to find it. When the hidden cache is found, the finder signs the log book and, if they wish, takes an item and returns the cache to its hiding place. Geocaching etiquette requires if you take an item you must leave an item. The finder then goes to the Web site where the coordinates of the cache were posted and posts his name as the most recent finder of the cache. If you think this sounds easy, think again. The coordinates, even if they are exact, will put you within 10 to 15 feet of the cache, but that's a circle with a radius of 10 to 15 feet. The cache could be anywhere within that large circle. Figure 6.1 is a photo of a typical cache.

Geocaching has developed into a worldwide activity with literally hundreds of thousands of caches hidden all over the world. If you enjoy the adventure of seeking and finding hidden treasure and like to roam in the great outdoors, geocaching with a GPS receiver can be a fun activity all in itself. Coupled with a ham radio flavor it can add another dimension to the traditional foxhunt. If you would like to know more about geocaching and look up the location of a hidden cache near you, log on to **www.geocaching.com**. At this site you can check for caches near your location and retrieve the coordinates of the one nearest you.

Figure 6.1 – The contents of a typical geocache. The geocache tradition is to take an object from the cache (a "prize" of sorts) while leaving another object for the next person. The notebook is for leaving a record of your discovery, somewhat like a guestbook at a wedding. After you exchange objects and scribble your comments, the cache should be re-sealed and carefully placed back where you found it.

# GEOCACHING WITH A HAM FLAVOR

This can be a great activity for an Amateur Radio Club that can provide its members practice in honing both their GPS skills and their radio direction finding capabilities. A group of hams are divided into teams of two to three people each, depending on the number of members available. Each team must have at least one GPS receiver and a handheld VHF or UHF transceiver. A reasonably large open space such as a park or a lightly wooded area is also needed. A day or two ahead of time a trusted member of the group hides several plastic containers that each contains the coordinates of a hidden low-wattage transmitter. The number of containers must equal the number of teams. The trusted member who hides the containers must carefully note the coordinates of each container as provided by his GPS receiver. A final container which contains the small low wattage transmitter will be hidden on the day of the hunt. The hidden transmitter is the ultimate goal that the teams must find. Each of the initial containers has two clues written on a log sheet: the coordinates of the hidden transmitter and its transmitting frequency. The UHF fox transmitter described in the March 2005 issue of *QST* Magazine "A Miniature UHF Fox Transmitter" by Dave Bowker, K1FK, is an excellent choice for the hidden transmitter. It is small, easy to build with few parts, and has a range of only ¼ to ⅓ mile. Figure 6.2 is a picture of Dave's UHF fox transmitter.

On the day of the hunt the trusted member hides the transmitter and tunes it to the desired frequency. The lowest amount of power available should be used. Each of the teams is given the container coordinates which they must successfully install in their GPS receiver. Each team must find their initial container to get the coordinates and frequency of the hidden transmitter. Once this is done they must then install the new coordinates into their GPS and start the hunt for the hidden transmitter. This all seems so easy, but it is more difficult than it sounds. First the initial coordinates must be installed correctly in the GPS, which will bring the teams within 10 to 15 feet of where the containers are hidden. They must then search for each container. Depending upon how well they are hidden, this could require some effort. Once the initial

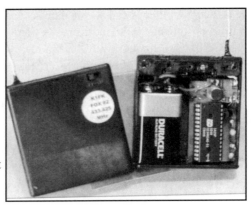

Figure 6.2 – Photo of UHF Fox Transmitter designed by Dave Bowker, K1FK.

container is found they must then install the new coordinates in their GPS and search for the hidden transmitter. Then, when they are in the immediate vicinity of where the transmitter is hidden, they can now use their handheld transceiver to find it. The winning team could be rewarded with a prize or an appropriate certificate.

This could be an exercise in how to install waypoints in a GPS and use the "GO TO" function of the GPS to learn the location of the containers. It is also an exercise in the use of the handheld transceiver in radio direction finding.

## GPS CHICKEN

Here's a little game you can play with your GPS while you are looking for the containers in the geocaching game. When you start out to look for the first container mark your starting position with the **MARK** button on the GPS and call it "Home". Now as you proceed to find the first container the GPS will record your track and distance covered. Arbitrarily pick a target distance, say ¼ mile. Then, while closing in on the hidden container, and without looking at the GPS, try to guess when you have reached the target distance. When you or your partner thinks you are there call "stop." Then use the GPS to check the distance. If the distance covered is over the target distance the caller wins. If the distance is under, the caller loses.

## FIELD DAY WITH GPS

Field Day probably gets more hams into the outdoors than any other Amateur Radio activity. This is an opportunity to set up and operate our stations as if we were participating in a real disaster recovery effort. But finding a desirable location to set-up operation, particularly in urban areas, can be difficult. Buildings can be a real obstacle and shopping malls and vehicle traffic are source of wide spectrum interference. Even in non-urban areas the challenges of surrounding hills and valleys can pose problems for VHF and UHF operation. This is where your GPS along with a topographic map can be of help in finding that great Field Day location.

As was discussed in Chapter 5, a topographic map shows changes in elevation with contour lines. With a topographic map you can find the high spots as well as the low areas of any particular location. In addition, topo maps will list objects such as prominent buildings, factories, vegetation, etc., using symbols to mark their location. Roads are also shown. This is important because you will need some way to access a chosen site.

Using the topo map you can search for the right spot for radio activities that will give you the advantages needed for the type of operation you plan for

Field Day. For example, if you are considering VHF or UHF operations from your Field Day location you will want to have the advantage of a high point from which to operate. You will also want to pick a location that is relatively free of building structures in the direction in which you wish to direct your signals. Since trees and tall vegetation can absorb and scatter VHF and UHF signals you will want to ensure these also are not an obstacle. If your Field Day operation includes HF operation, and most do, you will want to ensure that any tall hills and ridges are a significant distance away from your operating location so as to not interfere or have an adverse impact on propagation. In the *ARRL Antenna Book* Dean Straw, N6BV, has a revealing discussion on terrain and its influence on antenna elevation angles for HF operations. Dean provides a copy of his software, *High Frequency Terrain Analysis (HFTA)* with the *Antenna Book*. Using this software along with a topo map can greatly aid in determining the suitability of a location for the type of Amateur Radio activities you plan for Field Day.

So, where does GPS fit into all this? First, in the planning stages you can use the topo map to select several locations where it appears you would have the best chances for successful operation. Then, note their geographical location in latitude and longitude and mark these as waypoints in your GPS. Next, using the "GO TO" function of the GPS follow the directions given by the GPS to the exact location of each proposed site where they can be evaluated for suitability. Picking a location from a topo map is a wonderful idea, and it will give you a good overview of its suitability for your planned operations, but an actual on-site evaluation will allow you to determine if anything has changed since the map was published and to select the best of the proposed sites for your planned activities.

If your Field Day activities include reporting and exchanging coordinates with emergency and disaster control centers, the GPS can greatly aid in this regard as it will allow you to pinpoint exact coordinates to within 10 or 15 feet of any event. Having accurate position reports eliminates the guess work. This is especially true when street sign and markers have been blown over or destroyed in a disaster and there is little to use in identifying a location. The GPS has been successfully used by disaster response personnel over and over again to coordinate the deployment of recovery teams, to track and aid in reporting the location of victims, and to direct the response of fire, rescue and medical personnel where needed.

## AUTOMATIC POSITION REPORTING SYSTEM (APRS)

APRS is an Amateur Radio application utilizing packet radio in an unconnected mode of operation that was developed by Bob Bruninga,

WB4APR. Conventional packet radio operates in the connected mode; that is, one station communicating with another station in a point-to-point mode of operation. APRS allows multiple stations to communicate with one another on a real-time basis. In the words of Bob Bruninga, "APRS turns packet radio into a real-time tactical communications and display system for emergencies and public service applications and global communications." APRS uses Amateur Radio to transmit position reports, weather reports, and messages between multiple users, and when coupled with a GPS unit it can be used to track and display the position of a moving object as reported by an APRS station. APRS uses the "beacon" function of packet radio to accomplish all of these functions.

It is the content of the APRS packet that changes it from the ordinary beacon packet into an APRS packet. The APRS packet will contain the latitude and longitude (position) of the transmitting station, the station type (weather stations, home fixed stations, mobile stations, digipeaters, portables), and information in a specific APRS format that allows the receiving stations to process it and display it on a map showing the location of the transmitting station. If a station is mobile and in motion, the position of the station will change on the map as the position is updated either automatically with a connected GPS unit or manually with a connected notebook computer. Figure 6.3 shows the location of W4GNS in an 18-wheel truck traveling 65.5 miles per hour down highway I-381. In addition to tracking stations that are in

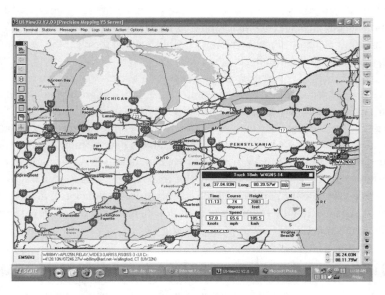

*Figure 6.3 – APRS map showing the location of W4GNS as he travels down Highway 381 in his 18-wheel truck.*

motion and automatically transmitting their locations, ARPS can be used to track objects such as hurricanes, fires, parades, marathons, etc, when the position of these is entered into the system by an APRS reporting station.

An APRS station can be set up in a fixed environment like a home or office, or a mobile environment like an automobile, boat, or airplane. There is even an APRS station on the International Space Station currently orbiting the Earth. All that is required is an Amateur Radio station, a packet radio terminal node controller (TNC), a computer and APRS software for the computer. Actually, with the right software, your sound card in the computer can be made to substitute for the TNC. If you have a transceiver that has built-in software and TNC ability like the Kenwood TM-D700A and the handheld TH-D7A(G) all you need is a GPS and a power source. Appendix B has a description of how to use a computer's sound card to substitute for the TNC. The standard connections that you would use to connect a packet radio station together also apply to APRS. When you add a GPS to the mix you will need a way to connect the GPS to the computer. If the computer has two serial ports use the second port for the GPS. If your computer has only one serial port you will need a single port switch cable (HSP) to connect to the computer and the two separate data ports. If you plan to go mobile with an appropriate transceiver like the Kenwood TM-D700A and a GPS unit then all you need for a connection is a cable from the data port of the GPS to the transceiver port. Figure 6.4 is a typical APRS map showing the locations of K1DAV, K1ARC. and N1FNE. Note the symbols that are used to describe each type of APRS station.

Since 1997 interfaces to the Internet have made APRS a global communications system for real-time traffic. Called *APRS-IS* (Automatic Position Reporting System – Internet Service), it is an Internet-based network which interconnects various APRS radio networks throughout the world, and even space, as the International Space

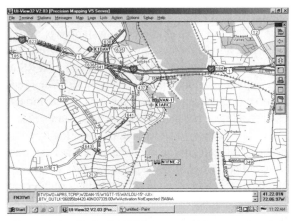

*Figure 6.4 – Typical APRS topographic map showing the locations of K1DAV, K1ARC, and N1FNE. This map uses the APRS symbols for a fixed station, a Red Cross station, and a mobile station.*

Station has an APRS digipeater in operation. Some APRS stations function as IGates (Internet Gateways) that receive data from the local area APRS stations and forward it to one of several Servers which in turn make it available to the World Wide Web. Steve Dimse, K4HG has a Web site called FINDU.com where anyone can find any APRS station that is transmitting and retrieve its transmission. Through the use of the Internet, APRS has become a real-time, world-wide activity.

Jutta Richter, DL1HXB, and Heiko Sauber, DO3SYM, have made good use of ham radio and APRS aboard their 100-foot sailboat the *SY MOMO*. They have sailed throughout Europe and North America using APRS and WinLink2000 to report their position so that friends and relatives could keep track of where they are at any time and know that they are safe while traveling around the world. You can see their current position by clicking on the link **www.findu.com/cgi-bin/find.cgi?call-DL1HXB**. The *SY MOMO* is well equipped with both marine and ham radio with marine VHF and HF/MF SSB, and VHF/UHF/HF ham radios including APRS and WinLink2000.

Many hams have used the tracking capability of APRS linked with GPS to report their position while traveling. Stan Horzepa, WA1LOU, and the author of *APRS Moving Hams on Radio and the Internet,* uses APRS to let his family know where he is located while traveling to the Dayton Hamvention every year. His family simply uses N1BQ's APRS Search Page **www.wulfden. org/APRSQuery.shtml** and asks for the last reported position of WA1LOU-8. No license is needed to access the APRS database; it is on the Internet for all to share and use.

Some hams have interfaced their home weather equipment to their APRS station to report the local weather conditions on a continuous basis in their area. All of these weather reporting stations show up on the map as blue circles with lines indicating wind speed and direction as well as other weather conditions. One unique application of using APRS to report weather conditions was implemented by Brian Wruble, W3BW, and reported in the June 2006 issue of *QST* magazine. Brian has a weekend home on the Sassafras River, an estuary on the Chesapeake Bay on Maryland's Eastern Shore, and has need to know the weather conditions at that location when he is away. To solve this problem he built a remotely located weather station at his weekend residence using a Peet Bros Ultimeter 2100 weather station with sensors for wind speed and direction, a rain gauge, and an outdoor temperature and humidity gauge. He connected the weather sensors to a Kantronics KPC3+ TNC and then fed the output of the TNC to an ICOM IC-2100H 2 meter transceiver. The known fixed coordinates of the weather station were then entered into the TNC. The TNC uses its normal GPS functionality instead to input data from the weather sensors and format the weather data for transmission by the ICOM transceiver as APRS packets. The APRS packets from Brian's weather station are picked up by an IGates station and entered into the Internet. At any place in the world Brian can find out what the weather conditions are at his weekend home by simply going on the Internet to the FINDU Web site and pulling up his location as reported by the FINDU data lookup and mapping service, which is an extension of APRS-IS. You can see a live presentation of active APRS stations world-wide at **www.aprs-is.net** and click on "Live APRS-IS" in the sidebar. Figure 6.5 is a picture of Brian's weather station.

In Onondaga County, NY there is a wilderness area where hikers are often lost. The disaster response personnel there make a unique use of APRS. Tracker radios linked to a GPS receiver are placed on search dogs using a special harness. Figure 6.6 is a picture of the Tiny Trak, similar to the units used with the search dogs. When the dogs are deployed in the rough mountain terrain, their movements are broadcast by the APRS tracker radios and observed at the Search and Rescue Center on a computer screen. Overlaying the tracking coordinates on a topographic map of the area gives the Center's personnel prior knowledge of the terrain where the dog is searching as well as the exact location. When the dog's movement stops for a period of time, indicating that a person has been found, the coordinates are recorded and a search and rescue team is dispatched to that area. The search and rescue team uses GPS receivers in which they have entered the coordinates as a waypoint where the dogs have indicated a person is located, and use this waypoint to guide them to the exact location. In effect, the use of APRS with GPS takes the

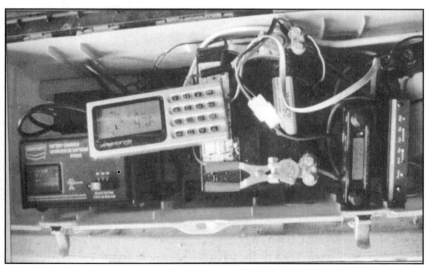

*Figure 6.5 – Brian Wruble's, W3BW, weather station showing the Peet Bros weather controller, 2-meter transceiver, and TNC.*

*Figure 6.6 – A photo of the Tiny Trak.*

"search" out of search and rescue, at least for the humans.

APRS has proven over and over again to be a very effective and useful tool in search and rescue operations and emergency operations management. The ability to know the exact location of victims and rescue teams can result in saving lives and is critical to team coordination while in the field. This type of emergency communications was put to the test during the hurricanes that hit the Gulf Coast states in 2005. The impact of Hurricane Katrina on communications networks caused enormous destruction and disrupted commercial communications for weeks. Three million telephone lines were blown away in Louisiana, Mississippi, and Alabama; thirty-eight 911 call centers were completely knocked out; over 1000 cell phone sites were completely washed away; and 100 commercial broadcast stations were knocked off the air. Nearly 1000 Amateur Radio Emergency Service (ARES) volunteers responded to the call for help and restored emergency communications in the area. They brought their emergency communications vans that provided not only voice communications but also data transmissions such as APRS. Using APRS linked with GPS, command center personnel successfully pinpointed rescue efforts and coordinated the movement of rescue personnel both on the ground and in the air

There is no doubt that the use of GPS linked with APRS has its best work with emergency communications. Imagine the possibilities of emergency communications vans equipped with APRs and computer equipment linked with GPS. As the various teams of weather spotters, search and rescue personnel, emergency medical personnel, and fire and police equipped with APRS trackers move through a community their positions are accurately tracked on a map displayed on the computers in the emergency communications vans. This information can be instantly relayed to a command post, or the emergency communications van itself can serve as the command post. Either way, on-scene commanders are aware at all times where each emergency response team is located. As the teams change locations their positions change on the display. All this is done without having to waste radio time informing the control station of their locations and coordinating their movements. Because the APRS network is by design a real-time tactical communications system, users of this system do not need to know in advance the calling stations or routing information such as nodes and repeaters. APRS provides universal connectivity to all stations at all times and avoids the limitations of a connected network. As the command post receives calls from various areas needing assistance they can plot the coordinates on the map and direct the nearest emergency response vehicles and teams to the area. As time is always critical in emergency situations, the ability to make decisions based on real-time knowledge could be a determining factor in providing life saving assistance.

The Brevard County Emergency Amateur Radio Service (BEARS) in Brevard County, Florida has an emergency communications van that is equipped with HF, VHF, UHF, and ham television equipment, as well as GMRS, CB, Marine, Fire, and Red Cross radios. There is also an APRS transceiver with GPS that is used to broadcast the location of the emergency communications van at all times when deployed. Weather data from the on-board weather system is also transmitted constantly through the APRS system and displayed on APRS maps at receiving stations. The driver of the van has a GPS receiver at his disposal to use in finding the coordinates where the van is directed to deploy. This is a real necessity when street signs and markers are no longer available to aid in finding exact locations. Support vehicles that work with the emergency communications van are also equipped with APRS and GPS receivers so that everyone knows in real time where all

Figure 6.7 – The BEARS1 Emergency Communications Van and members of the BEARS emergency Response Team.

*Figure 6.8 – The interior of the BEARS van showing the multiple operating positions.*

*Figure 6.9 – Rear view of the BEARS van showing the crank-up tower on the roof of the van and the mount where the tower is attached when deployed.*

the vehicles are located at any given moment. Figure 6.7 is a photo of the van and members of the BEARS team. Figure 6.8 shows the interior of the van and the multiple operating positions. Figure 6.9 is a photo of the van showing the crank-up tower that is mounted on the rear of the van when deployed. The BEARS emergency response team is well prepared to meet any emergency condition and is probably one of the best equipped and best trained emergency response units in the country.

We have looked at several applications where ham radio and GPS are congenial partners, from the fun and games we can have with these two technical marvels to the more serious side of emergency operations and management. One of the best and most popular applications of using GPS with ham radio is found with APRS, and in the next chapter we will take a look at how we can connect GPS with a ham radio transmitter/transceiver to make an APRS station.

# 7

# Making It Happen with APRS

Most hams have heard of APRS and know that is a form of packet communications, but relatively few have made APRS a part of their station or operation. Perhaps this is so because there is something mysterious about this mode of operation, and information on this subject is relatively hard to find. APRS is a great way to incorporate GPS into ham radio and in this chapter we will try to take some of the mystery out of APRS and provide some details on how to get started. Let's start first with a discussion of what APRS is and how we can make practical use of this mode of operation.

## WHAT IS APRS?

As discussed in the previous chapter, APRS is an acronym for Automatic Position Reporting System, a packet communications system that was developed by Bob Bruninga, WB4APR, for the rapid exchange of digital data. To quote Bob, "The Automatic Packet Reporting System was designed to support rapid, reliable exchange of information for local, tactical real time events or nets. The concept is that each station with new information transmits his new data to everyone in the net and every station captures that information for consistent and standard display to all participants." Although the original concept of APRS did not include vehicle tracking, but when GPS rapidly became affordable for the civilian user in the early '90s it was added to the system. Since then the tracking capabilities of APRS have been used in public service events such as parade monitoring, tracking the runners in marathon races, and in emergency and disaster control work, to name a few.

If you are familiar with packet radio, you are aware that in the traditional packet radio operations the communications take place on a one-on-one basis.

Basically two packet switching stations are connected exclusively to one another by a radio link using *connected* packets. Although APRS is a mode of packet switching operation, the difference is APRS uses *unconnected* packets to broadcast data to all stations that can receive the APRS packets, not to just one single station. APRS uses unconnected packets to permit any number of stations to send and receive data much like voice users would on a voice net. Any station that has data to send simply transmits it and all stations on frequency monitor and collect all data transmitted. Recognizing that one of the basics of real-time communications is the knowledge of where all stations are located, an APRS packet contains the station location in longitude and latitude, and the station type. There are several APRS station types, such as home station, portable station, mobile, digipeater, weather reporting station, and so forth. This information is in a specific digital format that allows receiving APRS stations to process the information and display an appropriate symbol on a map showing the location and type of the transmitting station. If the transmitting station is in motion while transmitting, the position of the station will change on the map display as new packet information is received. In addition to tracking moving stations, APRS can also track objects if their longitude and latitude are entered into the system. For example, many of the recent hurricanes in the Southern part of the United States were tracked by hams using APRS, providing a valuable service in weather reporting.

In 1997 APRS became a worldwide communications system using the Internet's World Wide Web. There are APRS stations that function as Internet gateways (IGates) that relay APRS packets to servers that are connected to the Internet. These servers collect APRS data from the local area and disseminate it internationally on a real time basis via a variety of Web pages. Anyone who has a Java capable Web browser can view these pages on their computer. In this case you don't need a ham radio or even have to be a ham to view selected APRS information. In fact, Steve Dimse, K4HG, maintains a Web site at **www.FINDU.com** which contains a data base of all APRS packets processed through IGates anywhere at any time on a up to the minute real-time basis.

It should be obvious to you now that APRS is an exciting ham radio activity that has possibilities limited only by the imagination of the user. Let's now explore the equipment needed and the different methods of connectivity available depending on how we intend to set up an APRS station.

## SYSTEM REQUIREMENTS FOR APRS

**The Fixed Station Configuration.** The frequency of choice for APRS operation throughout the North American continent is the 2-meter frequency of 144.39 MHz. There are also some APRS stations on the UHF frequency of 445.925 MHz, but this is not the primary frequency for APRS work in North America. Other countries use different frequencies. The basis of an APRS station in North America then, is a 2-meter transmitter and receiver (or transceiver) and appropriate antenna. If your planned operation will be a fixed station, you will next need a terminal-node controller (TNC). The TNC is a device that takes the data (digital information) from your computer or data terminal and assembles it into packets. It then converts the data to audio tones (analog information) that are passed to your transmitter or transceiver. It also receives the audio tones from your receiver and translates them into data that your computer or terminal can understand.

To generate the data that is fed to the TNC you will also need a computer or appropriate data terminal. In general, any modern computer will suffice to operate with APRS software with the only concern being the amount of RAM and hard disk storage space available. Maps take up a lot of hard disk storage, and depending upon the number of maps you intend to use you may need to ensure there is enough storage space on your hard disk to meet your requirement. Newer home computer systems tend to come with tens of gigabits of hard disk storage which should provide more than enough storage space. There are several TNCs available that are suitable for use with APRS. Some of the more popular TNCs are models of Kantronics, PacComm, MFJ and Timewave. Because a fixed station does not change position, it does not need a GPS receiver; however, when the APRS software is initially configured a GPS receiver may be used to determine the station's location in longitude and latitude. A city or town, and street address are not adequate to configure the software. You will need the longitude and latitude coordinates. Although not required for a fixed station, the easy way to get this is with a GPS receiver, but if one is not available you can use a topographic map like that discussed in Chapter 6. The United States Geological Survey (USGS) topo maps are perfect for determining coordinates.

Next you will need APRS software to run on your computer. There is a variety of APRS software available depending on your computer's operating system. The following is a listing of available APRS software by the more popular computer operating systems (OS):

> *Windows* OS (16 bit)--*UI-View* and *WinAPRS*
> *Windows* OS (32 bit)--*APRS+SA*, *APRSPoint*, *UI-View* and *WinAPRS*
> *Mac* OS-MacAPRS and XASTIR--(*XASTIR* runs under *X-Windows*, which is included in the Mac OS
> *Linux* OS--*X-APRS* and *XASTIR*
> *Unix* OS--*XASTIR*

All of the above listed software can be downloaded from the Internet. The following are the Web site addresses for the listed software:

*UI-View* - **www.ui-view.org/**
*WinAPRS* - **www.winaprs.org/**
*APRS+SA* - **www.tapr.org/~kh2z/aprsplus/**
*APRSPoint* - **www.aprspoint.com**
*MacAPRS* - **www.winaprs.org/**
*X-APRS* - **www.tapr.org/software_library.php?dir=/aprssig/**
    **winstuff/xaprs**
*XASTIR* - **www.XASTIR.org/**

*UI-View* (16- or 32-bit) was written by the late Roger Barker, G4IDE. The 32-bit version supports WA8DED/TF host mode. Host mode means that *UI-View* can be used with an extremely wide range of packet hardware. *WinAPRS* (16- or 32-bit), *MacAPRS*, and *X-APRS* were written by Keith Sproul, WU2Z and Mark Sproul, KB2ICL. *WinAPRS* as a minimum will support a PC with a 386 processor running the *Windows 95* operating system. *MacAPRS* is a Mac version of APRS, and *X-APRS* is a *Linux* version of APRS. *APRS+SA* was written by Brent Hildebrand, KH2Z, and uses *Delorme Street Atlas USA* software for its maps. Delorme can be reached at **www.delorme.com**.

The most popular APRS software today is *UI-View*. It supports a wide variety of maps, and technical support is provided from the Web site **www.peak-support.com**. This Web site provides the beginner in APRS everything needed to get started, including links for maps, software, available applications, and technical support.

Let's next consider how we can connect all of this equipment together

*A typical DB-25 male connector.*

*Two popular serial connectors. Top: DB-9 female. Bottom: DB-9 male.*

to make an APRS station. At the radio port, TNCs typically use a female DB-9 or 5-pin DIN connector to provide the audio in, audio out, and push-to-talk (PTT) connection that will go to your transceiver. Connect the audio out of the TNC to the audio in of your transceiver. This is typically the microphone input (MIC) connection on most transceivers, but some have separate inputs for AFSK tones that are labeled "AFSK In". If this is the case with your transceiver, use this connection rather than the MIC In to allow you to use the voice mode without having to take down the TNC connection. The AFSK input is also more suited to data operation than the analog mode provided by the microphone input.

Now connect the audio in of the TNC to the audio output of the transceiver. Typically this is the speaker or headphone connector on the transceiver. Some models in addition to having an AFSK In connector also have an AFSK Out connector. Use this connection if it is available. Connect the PTT line of the TNC to the PTT connection on the transceiver or transmitter. This is usually found on the microphone connector, but some models also have a separate PTT connection as well. If available, use this connection. The last connection is to connect the TNC ground connection to the ground connection on your transceiver. Figure 7.1 is a drawing showing the typical connections between the TNC and the transceiver.

The above described connections assume that a 2-meter base station will be used as the station radio rather a handheld transceiver. Most handhelds use a common connector for audio input and PTT. Connecting the TNC audio input and PTT line directly to such a handheld will not work. You will need to typically add a resistor of 1kΩ to 2 kΩ and a .01 to 1.0 μF capacitor in the line to make it work. Figure 7.2 shows this connection. Values of the resistor and capacitor can vary depending on the model of the radio. You will need to consult the owner's manual to obtain the correct value.

Now, our last task is connecting the TNC to our computer. Most TNCs have a serial port for this connection that uses a female 25-pin DB-25 connector. Although this connector is capable of handling 25 signals,

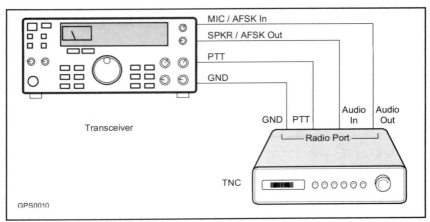

Figure 7.1 – A typical 4-wire connection between a transceiver and a TNC.

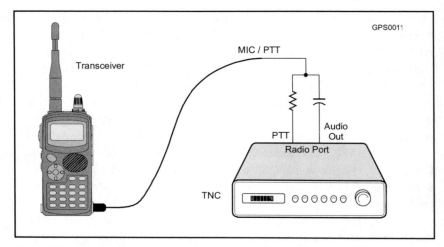

*Figure 7.2 – A typical MIC/PTT connection on a handheld transceiver to a TNC where a resistor and a capacitor are used to isolate the signals at the TNC.*

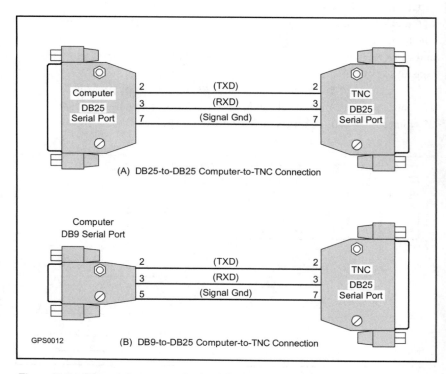

*Figure 7.3 – The minimum required cabling between a TNC and a computer.*

the TNC needs only three: transmit data, receive data, and signal ground. Most computers have a serial port that uses a male DB-25 or male 9-pin DB-9 connector. If your computer has a 25-pin serial port it uses the DB-25 connector and the wiring between the TNC and the computer will be as follows: pin 2 is transmit data, pin 3 is receive data, and pin 7 is signal ground of both the TNC and computer connector. Figure 7.3 shows this connection. If your computer has a 9-pin serial port, it uses a DB-9 connector and the wiring will be as follows: pin 2 is transmit data of both the TNC and the computer, pin 3 is receive data of both the TNC and computer, and pin 7 of the TNC will become pin 5 at the computer connector. If your computer has a Universal Serial Bus (USB), you can avoid the trouble of making up your own serial connector cable by purchasing a serial cable that has a USB connector on one end and a DB-25 connector on the other. Most computer and electronic stores carry this type of cable.

Now let's assume we have our TNC connected to our transceiver and the computer and we have downloaded the appropriate APRS software for our computer. The only thing left to do is to install the software. There are so many versions of APRS software, depending upon the specific computer and operating system, that it would be impossible to cover them all in this chapter. Most have detailed instructions for installation that make it easy to follow. What are covered here are those generic terms that are contained in all APRS software, and that require some action on your part. These are the minimum parameters that you must set to properly operate APRS.

**Call Sign and SSID:** Configure the software with your Amateur Radio call sign and a secondary station identifier (SSID) if one is desired. The SSID follows the radio station call sign and is used to differentiate between two or more APRS stations operating under the same call sign. You can have up to 15 SSIDs, but if you choose not to specify a SSID the default will be zero. For example you can have a fixed station with the call sign W4WCF-1 and a mobile station with the call sign W4WCF-2.

**Alias:** This parameter indicates whether your APRS station is RELAY or WIDE. RELAY and WIDE determines the digipeater path your station will use to be received by other APRS stations. If your path is set incorrectly, then only the stations that can hear your station directly will receive your transmission, and the stations beyond the range of your transmissions won't receive any of your transmissions. By setting your alias to RELAY any APRS RELAY station that receives your initial transmission will retransmit it. Since the principle of APRS is the rapid dissemination of real-time information, once the initial APRS RELAY station retransmits your packets, they will not be repeated by other APRS RELAY stations that might also receive your transmission. To provide even wider dissemination of your transmission you could use the alias of WIDE making your path RELAY WIDE. If a WIDE digipeater station received your transmission via the APRS RELAY station it would be retransmitted by its digipeater.

You can choose to ignore the digipeater path setting when you configure your software and let the software use its default path. If you are not aware of your local APRS network, this is the best way to get your station on the air. Once you become knowledgeable of your local APRS network you can then fine tune your digipeater path setting.

**Symbol:** You are required to select a symbol that represents your station on the APRS map. As a fixed station this probably will be the symbol "y" for house with Yagi antenna or the symbol "-" for house with a vertical antenna. Appendix A lists all the symbols used with APRS software.

**Position:** This is a critical parameter that includes the longitude and latitude coordinates of your station. This requirement was discussed earlier in this chapter.

**Position Text:** This can be your name, address, or whatever you want it to be. It is a short string of text that the APRS software sends with your position packet information. There is a timer associated with your position packet that you can adjust. For a fixed station it is recommended the timer be set to 30 minutes. For a mobile station the timer should be set to every 1 or 2 minutes.

**Status Text:** This is a short string of text that the software sends whenever it transmits your beacon. It can be your name, location, or whatever you wish it to be. There is a timer associated with this application also that you can set to determine how often you would like your beacon to be transmitted.

**Port:** This parameter indicates which computer port you used to connect to the TNC and any other equipment used in your APRS station.

**Baud Rate:** This is the baud rate between your computer and the TNC. For APRS this is normally 1200 bps.

**Data Bits:** This is the number of data bits that represent an alphanumeric or control character used between your computer and the TNC. You normally can find this information in the user manual that comes with the TNC.

**Stop Bits:** This is the number of data bits that follow a character to indicate its end. The choice may be 1, 1.5, or 2. Consult your TNC user manual.

**Handshaking Protocol:** This is the protocol used for coordinating communications between your computer and the TNC. If unknown, select "none".

**TNC:** This is the brand and model number of your TNC that is connected to the computer and whether it is dual band or single band. Again, consult your TNC user manual for this information.

**RF Equipment:** This is where you enter your transmitter power, the antenna height above average terrain, and the antenna gain. The transmitter power is in watts and the antenna gain is in dB. If your transmitter power is above 81W simply enter 81. This is the default power level to promote the use of minimum power in APRS networks.

The antenna height is the height above average terrain, not height above sea level. This is the average height of the antenna as it relates to the average height of the terrain in a 10 mile radius of your antenna. For example if your antenna is 500 feet above sea level and the average height of the terrain in a 10 mile radius is 450 feet, then your height above average terrain is 50 feet.

To determine the average height of terrain in a ten mile radius it is best to use a topographical map of your local area and record the height of the terrain in 2-mile increments along the compass directions of N, NE, E, SE, S, SW, W and NW for 10 miles out from your location. When completed, you should have 40 points recorded. Add them together and divide by 40. Take the result of this calculation and subtract it from you antenna height above sea level. The difference is your height above average terrain.

**UTC Offset:** This is where you enter the number of hours plus or minus that your local time differs from Greenwich Mean Time which is now called Coordinated Universal Time.

**Registration Number:** You are encouraged to register your APRS software after you download it from the Internet. Most APRS software is Shareware which comes with the understanding that you can try out the software for free to determine its suitability, but if you decide to keep it you are obligated to pay for and register it. If you have registered your software, this is where you will type in the registration number. By doing so, you save having to reconfigure your software every time you start APRS.

**Heading and Speed:** This parameter applies to a mobile configuration of APRS and can be implemented by connecting a GPS to the transceiver. This will be discussed in the next section.

Having connected your TNC and computer to your transceiver, and downloaded and installed the APRS software, you are now ready to put your APRS fixed station on the air. All the operation options available for your APRS station are not covered here. The best book on this subject is *APRS Moving Hams on Radio and the Internet* by Stan Horzepa, WA1LOU, and is available from the ARRL.

## THE MOBILE AND PORTABLE APRS STATION CONFIGURATION

Let's look now at the system requirements for a portable or mobile APRS station. This is where we can make good use of what we have learned about the GPS because this type of APRS station configuration really needs the positioning capabilities that the GPS provides. Mobile means that your station

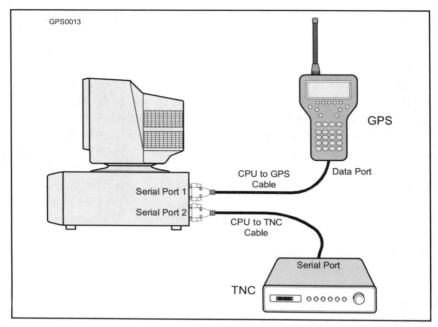

Figure 7.4 – Typical connection between a GPS, TNC, and a computer where the computer has two serial ports.

generally is in motion and constantly changing position. The GPS has the capability to keep up with every position change on a minute-by-minute basis and accurately report the new positions. By connecting a GPS receiver to your mobile station this function can be done automatically, and as your position changes it will appear accurately on an APRS map.

Portable implies your station is not at your normal fixed location, and could be in motion or at another fixed location. In either case you would not probably be aware of the longitude and latitude of where your portable station was located, but the GPS would know and accurately report it.

In the mobile/portable configuration we have multiple choices of how we can connect up our station and the type of equipment we can use. First, we can set-up our station with a TNC connected to the transceiver as in the fixed station operation. Next we could add a laptop computer and a GPS. The connections required for this configuration would be as shown in Figure 7.4 or 7.5. In Figure 7.4 we are using a computer that has two serial ports. In this case the laptop computer is connected to the TNC through one serial port and the GPS is connected to the computer through the second serial port. The only complication with this connection is with the GPS receiver. Different GPS receivers use different connectors to connect to other electronic equipment. It is best to obtain this connector and cable from the GPS receiver manufacturer, or you can make up this cable using the information in your GPS receiver's user manual to obtain the proper pin connections. The GPS will have a data out, a data in, and a ground connection. At the computer end of the cable typically the pins 2, 3 and 7 are used for the serial connection as described in the TNC connection to the computer.

In Figure 7.5 we are using a laptop computer that has only one serial port. In this case you need to use a hardware single port switch (HSP) to connect both the TNC and the GPS receiver to the computer. A HSP switch is available from a variety of sources such as Kantronics, PacComm, and MFJ. The HSP switch is connected to the laptop computer using the computer's serial port and the cables from the TNC and GPS receiver are connected to the HSP switch as shown in Figure 7.5. In both types of connections (Figures 7.4 and 7.5) you must configure the APRS software so it is aware of these connections.

There are several mobile and handheld transceivers available that have a built-in TNC and APRS software installed. These can be a standalone APRS station. The only other equipment needed is a GPS receiver to make these transceivers a complete mobile or portable APRS automatic reporting station. The Kenwood handheld TH-D7A(G) and the Kenwood TM-D700A Dual Band Data Communicator are examples of transceivers that can be standalone APRS stations. These transceivers make APRS mobile or portable operation easy without additional equipment or software needed other than a GPS receiver.

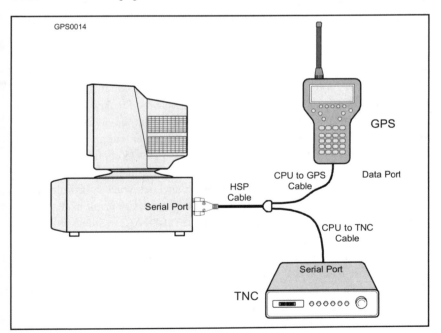

*Figure 7.5 – Typical connection between a GPS, TNC and a computer where the computer has a single serial port and single port switch (HSP) is used.*

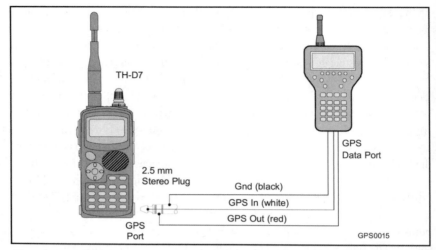

*Figure 7.6 – Connection between a GPS and the Kenwood TH–D7A(G) using the Kenwood supplied cable.*

Here's how you can connect a GPS receiver to the Kenwood TH-D7A(G). Kenwood supplies a cable with a 2.5 mm 3-conductor plug with this unit. Use this cable and connect the red wire in the cable to the GPS receiver's data out connector. Next connect the white wire to the GPS receiver's data in connector, and connect the shield to the GPS receiver's ground connector. This connection is illustrated in Figure 7.6. That's it. Using the software that is programmed in the TH-D7(A/G) you now have a functional mobile/portable APRS station. As an alternative, you can install the version of software you installed in the computer of your fixed station if you so desire. Kenwood makes this easy with an optional cable (PG-4W) that has the 3-conductor plug on one end to mate to the handheld and a DB-9 serial connector on the other end to mate to your computer's serial port. If your computer has only a DB-25 connector or a USB serial port, you will have to use an adapter which is available at most computer and electronic stores. Once the two units are connected, you can download the APRS software from the computer to the handheld. Connections to the Kenwood TM-D700A are similar to that of the TH-D7A(G) with a similar cable that comes with the transceiver to connect to a GPS receiver.

The possibilities of mobile operation with APRS are limitless. To quote Steve Dimse, K4HG, in the Afterword of *APRS Moving Hams on Radio and the Internet,* "So far, the focus of APRS has been on getting information from mobile users to central sites, whether a command post, EOC, base station, or the Internet and **www.FindU.com**. A big change I see coming is more focus on getting various forms of information to the mobile user. The possibilities are legion..."

# 8

# Selecting a GPS Receiver

Now that you know the basics of GPS and some of the things you can do with it, you may be ready to invest a few dollars in a GPS receiver. But, before you stroll over to the electronic store and start to browse around there are some things you need to know to make an intelligent purchase. Different tasks demand different kinds of GPS receivers, and while they all perform the function of providing position information it's what the receiver does with that information and how it is presented that varies from receiver to receiver. To make an intelligent purchase you need to first identify your goals for the use of GPS. Once you know what you want the GPS to do for you, the selection process becomes easy.

## IDENTIFYING ONE'S GPS GOALS

Ask yourself a few questions. Do I want to use my GPS receiver on land, on the water, or both? Or, am I a private pilot and would like to have my GPS with me when I go flying? What's my budget, how much can I afford to invest? These are the kind of questions you need to consider in determining the kind of GPS receiver to buy because there are receivers that are designed specifically for each of these applications.

If you own a boat you most certainly will want to have a GPS on board. The marine user should consider a GPS with a marine database. This type of GPS will include markers, buoys, and aids to navigation that apply to water navigation. But, do you want one that is fixed mounted to the boat or one that is handheld? Typically, fixed mounted units will be used only on the boat, and if this is your choice, you should select one that includes chartplotting. Handheld units are available with and without a mapping capability. If this is your choice, the mapping feature is only slightly higher in price and is well worth the extra cost. Other considerations for the mariner include:

- Will the GPS be operated in the direct sunlight?
- Will the GPS be operated on internal batteries or from the boat's power source?
- Is the receiver's screen large enough to read in choppy waters?
- Will I need an external antenna or can I use the one built in the unit?
- Is the receiver sufficiently waterproof to be operated in an environment exposed to sea and rain?

If you are a private pilot, a GPS receiver is a necessity. Of the more than 5300 public airports in the United States only 300 have the radio-beacon systems necessary to guide pilots to runways in poor weather conditions. With a WAAS (Wide Area Augmentation System) enabled GPS receiver a pilot can find the runway of any airport in the country regardless of the weather. GPS receivers made specifically for the aeronautical environment allow pilots to obtain the same quality of flight information as a traditional instrument landing system (ILS). Like the GPS receivers made specifically for the marine environment, those made for aeronautical use come in two varieties: panel mounted and portable models that can be moved from airplane to airplane. The fixed mounted models are generally connected to the other instrumentation in the aircraft and are therefore more dependable with the advantage of offering more flight information to the pilot. Some of the features of the aeronautical GPS include:

**Altimeter.** Most are accurate to within 5 to 10 feet.

**Alarms.** Most will provide an alarm when you near your destination, veer off course or approach a set waypoint. Most have a built-in Jeppesen database that allows you to view your flight course displayed over terrain maps.

**Nearest Airport.** Most have a database of US airports that you can select for emergency "nearest" option.

**Position Awareness.** Plots your position on a moving map allowing you to view your position even on the runway and taxiways.

If you spend a lot of time traveling in your automobile, a GPS receiver designed specifically for the automobile can be a great aid in finding a location whether it is a street, a gas station, a restaurant, or the best route to take on a long journey. GPS receivers designed for automobiles have a built-in database that contains an astounding amount of information. Most automobile manufacturers include a built-in GPS receiver as an optional equipment accessory. These usually have a wide screen that can be easily viewed from the driving position. Unfortunately, they also usually come with a high price tag, but there are GPS receivers made exclusively for the automobile that can be added on for a modest price. These include the Garmin Street Pilot, the

*A Garmin Street Pilot 7200.*

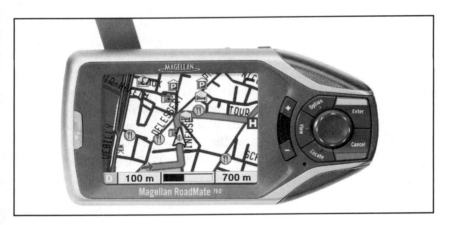

*This is a compact Magellan RoadMate.*

Magellan RoadMate, the Tom Tom Portable, the TeleType WorldNav, and the NavMan Navigator to name a few. Some of these offer voice prompts to aid in finding a particular location or street. Some come with touch screen capability so that you can give commands to the GPS simply by touching a place on the screen. All come with a database of practically every street, road, and highway in the United States. Most also include in their database the location of popular restaurants, major gas stations, major lodging establishments and points of interest such as parks, museums and zoos. The amount of information contained in the database of these specially configured GPS receivers is truly mind boggling.

The popular Garmin eTrex GPS receiver.

The tiny Garmin Geko.

If you want just a GPS receiver that you can use for a variety of purposes with no specific task in mind then you might be interested in a general purpose handheld GPS. These receivers are characterized by being small, portable, and battery powered with a built-in display. Some of these receivers have the ability to display nautical charts, topographical maps, or road maps from data cards inserted into the receiver. The antenna is built in the unit and some receivers allow it to be detached for mounting outside a vehicle. There are literally hundreds of models available that fit this purpose. Selecting one is influenced primarily by the company's reputation, support, and warranty of their product. Price and ease of operation also play a role in the selection process, but the competition between manufacturers of these types of GPS receivers has driven prices down to where a relatively high performance unit can be obtained for a modest price.

Most of us will want to own at least one general purpose handheld GPS even if we purchase one designed for our boat, car or airplane. The practicality of this type of unit lends itself so well to any type of activity where positioning information is needed. It is portable, carries its own power source, has the capability to display all types of information whether it is chart or maps, and is small and lightweight enough to be easily carried in our pocket.

## WHAT FEATURES ARE NEEDED?

After you have established your goals and determined what it is you want the GPS to do, you next need to determine what features are needed on the receiver to accomplish the tasks. Although all GPS receivers perform the basic task of providing positioning information, that is, providing your location in latitude and longitude, there are many additional tasks a GPS receiver can perform depending on the type and model you select. Perhaps the first additional feature you need to consider is whether the GPS receiver you choose should have a mapping capability. Today all GPS

A screenshot of the AvMap GeoSat 4T in action. This is the only in-vehicle navigator designed with amateur APRS in mind. For APRS operation, the included cable connects between the Kenwood TH-D7A or TM-D700 transceiver GPS port and into the GeoSat TMC port. See **www.geosat.us/**.

receivers except those models that are the most basic handheld units feature an electronic mapping capability. If you want to get only position fixes and navigate using traditional paper charts or maps, a receiver like the Garmin GPS 72, GPS 76 or Geko 201 will provide that capability and let you establish waypoints and plan routes. These basic GPS receivers are very reasonably priced and are very reliable. If your interest is in seeing your position on a detailed electronic chart or map then you should consider one of the GPS receivers that offer more than just basic positioning and waypoint management capability. Beyond the mapping feature, there are other features that should be considered when selecting a GPS receiver. Some of these are:

**Handheld, mountable or fixed mount.** For hiking, camping, small boats and general portability one of the pocket-sized handheld receivers is ideal. Almost all come with a color display that can be read in direct sunlight, and they all contain their own battery power source. They are modestly priced and most come with a preloaded base map with varying levels of detail. Some have the capability to add additional cartography with cartridges or chips that are inserted in a port on the receiver.

The next step up in size is the mountable GPS receiver. These receivers combine the properties of both handheld and fixed mount units. These receivers have larger displays than the handhelds, larger key pads, and can be powered by either their internal batteries or an external 12V dc source. They come with mounts that can be used on a vehicle dash or a boat helm, and can use either a built-in antenna or connect to an external antenna for better reception of the satellites' signal. Most come with mapping software that can be used for both highway and boating navigation.

The largest GPS units are the fixed mount GPS receivers. Most have high resolution displays and larger key pads that are most convenient to use. They have no internal power source and require a 12V power source from the vehicle where they are mounted. Most also require an external antenna.

**Cartography.** Although all modern GPS receivers, except the basic handheld units, have an electronic mapping capability and come with a preloaded base map, to achieve a full mapping detail with charts or maps showing a level of detail down to a few feet most use electronic moving maps designed for marine, automotive, off-road or back country applications that show your position directly on the map or chart. There are several ways this detail cartography is delivered.

**Preloaded.** This is a convenient feature that provides the cartography preloaded on the receiver using either flash memory or a hard drive. The GPS receivers that are specially designed for automotive and aeronautical use have this type of cartography.

**CD ROM Downloads.** Some receivers require that you connect them to a computer and download sections of a CD either directly to the receiver or to a data card that is later inserted in a port on the receiver. In some cases buying the CD allows you to download one area or region, and if you wish additional regions you will need to buy unlock codes for every additional region you choose.

**NEMA Data Output.** If you are interested in using the receiver for APRS, you'll need one that has a NEMA data output port to connect to the APRS TNC (Terminal Node Controller).

**Cartridges.** Many charts and maps are sold on small cartridges containing a regional data file that are inserted into ports on the receiver. If you wish data files on a new location you have to purchase additional chips or cartridges. The new Garmin handheld units use tiny MicroSD cards which are about ¼ the size of postage stamp. These tiny cartridges contain up to 2GB of storage.

Competition in the GPS receiver marketplace has brought a continued offering of the latest technology and model variety from a host of companies. At the same time that technology and variety has increased, prices for these developments have decreased. It is possible to purchase a basic handheld GPS receiver for $100 or less today, and it will perform better and have more capability than the $40,000 model of the original GPS receiver that was developed for the US Government.

This chapter has given you a brief overview of the types and variety of GPS receivers that are available today. One final word: After you have decided what you want to do with GPS and have in mind the type of GPS receiver you want, visit your electronic store and get the current information on the latest models that fit the category or activity you have in mind. Most dealers will let you handle their display models. This is a good way to get familiar with how particular units operate and determine if they really are what you want in a GPS receiver.

# Glossary

**2-D mode** – A two dimensional position fix that includes only horizontal coordinates (latitude and longitude). It requires adequate signals from a minimum of three visible satellites.

**2drms** – Two degrees of root mean square; a statistical statement that includes 95 percent of the samples

**3-D mode** – A three-dimensional position fix that includes horizontal coordinates (latitude and longitude) plus elevation. It requires adequate signals from a minimum of four visible satellites.

**accuracy** – The level of match between the GPS measured position, time, and/or velocity and its true position, time, and velocity.

**acquisition time** – The amount of time required for a GPS unit to lock onto three satellites to provide a 2-D fix of present position.

**almanac data** – Part of the navigation message transmitted by each GPS satellite that contains information on the orbits and status (health) of each GPS satellite.

**atomic clock** – A precise clock that operates using the elements cesium or rubidium. A cesium clock has an error rate of one second per million years. GPS satellites contain multiple cesium or rubidium atomic clocks.

**bearing** – Horizontal direction of an object from an observer, expressed as an angle from a reference direction, e.g., compass bearing, true bearing, relative bearing.

**cartography** – The art or technique of making maps or charts. Many GPS receivers have detailed mapping or cartography capabilities.

**chart** – A graphic representation of a part of the earth's surface transferred to a flat surface.

**chart chips** – Chart cartography used in most mapping GPS receivers that is provided by removable module usually referred to as chart chips or cartridges.

**clock error** – The amount by which the internal clock of an electronic device differs from the standard

**course acquisition (CA) code** – The standard positioning signal the GPS satellite transmits to the civilian user. It contains the information the GPS receiver uses to fix its position and time.

**cold start** – The power-on starting sequence of a GPS receiver that does not contain current almanac data.

**control segment** – The worldwide chain of control and monitoring stations that manage the GPS space segment.

**Coordinated Universal Time** – Replaced Greenwich Mean Time (GMT) as the world standard time in 1986. It is based on atomic measurements rather than the rotation of the Earth.

**course** – The direction that is to be or was traveled.

**course made good** – The resultant direction from a point of departure to a point of arrival

**course over ground** – Actual direction of travel of a vehicle over ground. In nautical terms this is the actual travel of a boat over the sea bottom.

**cross track error** – The distance off the desired course.

**datum** – A mathematical representation of the irregular, not quite spherical, shape of the Earth; defines latitude and longitude for the area covered.

**deviation** – The numerical difference in degrees, measured east or west, between the magnetic value and the compass value of a given direction.

**Differential Global Positioning System (DGPS)** – A technique used to improve the accuracy of the GPS using land-based medium frequency transmitting equipment.

**distance** – The spatial separation of two points and the length of the line joining them.

**dilution of precision (DOP)** – A measure of satellite geometry. The smaller the number, the better the geometry.

**ellipsoid** – A geometric surface whose plane sections are either ellipses or circles.

**equator** – An imaginary plane that passes through the center of the Earth perpendicular to the axis of rotation.

**ephemeris** – A collection of tables or data showing the position of the planets or heavenly bodies for every day of a given period; also an astronomical almanac containing such tables.

**fix** – A relatively accurate position determined without reference to any former position. Usually determined by nearness to a known carted object or by crossed (intersected) lines of position.

**geoid** – The particular equipotential surface that coincides with mean sea level and that can be imagined to extend through the continent

**Global Position System (GPS)** – A satellite-based worldwide navigation system using simultaneous signals from three or more satellites to establish highly accurate positioning.

**heading** – The direction in which a vehicle is pointed or a person is moving.

**hertz (Hz)** – The unit of frequency of an alternating current. One hertz equals one cycle per second.

**ionosphere** – The band of charged electrons 80 to 120 miles above the earth's surface.

**kHz** – Kilohertz or 1,000 hertz.

**L1** – The GPS standard positioning service signal that is transmitted in the clear on 1,575.42 MHz

**L2** – The GPS precise position service that is encrypted and transmitted on 1,227.6 MHz.

**L-band** – A portion of the microwave spectrum around 1,200 MHz.

**latitude** – Angular distance on the earth's surface measured north and south from the Earth's equator.

**longitude** – Angular distance on the Earth's surface measured east or west from the prime meridian extending through the observatory at Greenwich, England.

**MF** – medium frequency. The portion of the radio spectrum between 300 kHz and 3 MHz.

**MHz** – Megahertz or 1,000,000 hertz.

**nanosecond** – One-billionth of a second, or one thousandth of a microsecond of time.

**nautical mile** – A unit of distance equal to one minute of latitude and equal to approximately 6076.1 feet or 1.15 statue miles.

**NAVSTAR** – The official name of the GPS. An acronym that stands for *NAV*igation *S*atellite *T*iming *A*nd *R*anging.

**NMEA 0183** – The protocol establish by the National Marine Electronic Association that is used to exchange information with electronic equipment.

**prime meridian** – The meridian that passes through Greenwich, England. The prime meridian is measured at 000 degrees longitude.

**pseudorandom noise (PRN)** – A sequence of digital 1s and 0s that appear to be randomly distributed like noise but that can be reproduced exactly.

**radio direction finder** – A radio receiver equipped with a compass rose and a loop or other type of directional antenna to determine the direction of the source of a received radio signal.

**route** – A sequence of waypoints that combine to mark a proposed route of travel.

**selective availability (SA)** – A means of intentionally degrading the accuracy of the GPS standard positioning service signal.

**sky view** – A two-dimensional representation that shows the relative bearing and elevation angle of GPS satellites currently in view of a GPS receiver.

**space segment** – The constellation of GPS orbiting satellites.

**speed** – Rate of motion.

**speed of advance** – Intended or expected speed along a track.

**speed over the ground (SOG)** – Actual speed achieved relative to the ground.

**Standard Positioning Service (SPS)** – The normal civilian positioning accuracy obtained by using the single frequency L1 signal.

**T-Hunting** – Searching for a hidden transmitter by Amateur Radio operators, using a hidden VHF or UHF transmitter; sometimes called "Fox Hunting".

**true north** – The true direction to the North Pole.

**UHF** – **ultra high frequency.** The portion of the electromagnetic spectrum from 300 to 3,000 MHz.

**user segment** – The GPS receiver and its operator

**velocity made good** – Speed toward destination or final waypoint.

**VHF** – **very high frequency.** The portion of the electromagnetic spectrum from 30 to 300 MHz

**warm start** – The power-on starting sequence of a GPS receiver that contains current almanac data in its internal memory.

**waypoint** – A defined point or a latitude and longitude position entered in a GPS receiver.

**WGS-84 (World Geodetic System 1984)** – The mathematical ellipsoid used by GPS since January 1987. Most current charts and maps use this datum.

**Wide Area Augmentation System (WAAS)** – A US Federal Aviation Authority system to supplement GPS accuracy using geostationary satellites to transmit correction data on the GPS L1 frequency.

**World Geodetic System** – A consistent set of parameters describing the size and shape of the Earth.

# Resources

The best resource for information on GPS is the Internet. There are multiple categories of websites that offer information on just about any aspect of this marvelous location awareness tool. Perhaps the greatest challenge is in interpreting the wealth of information that is available using a simple search engine and typing in "GPS". To help organize your search we offer this list of resources by category starting with a listing of current GPS equipment manufacturers.

## GPS EQUIPMENT MANUFACTURERS

Eagle Electronics, PO Box 669, Catoosa, OK 74015
**www.eaglegps.com**

Furuno USA, Inc., 4400 NW Pacific Rim Road, Camas, WA 98607
**www.furuno.com**

Garmin International, 9875 Widmer Road, Lenexa, KS 66215
**www.garmin.com**

Humminbird, 108 Maple Lane, Eufaula, AL 36027
**www.humminbird.com**

Interphase Technologies, 2880 Research Park Drive #140, Sequel, CA 95073
**www.interphase-tech.com**

Lowrance, 12000 East Skelly Drive, Tulsa, OK 74128
**www.lowrance.com**

Magellan, 960 Overland Court, San Dimas, CA 91773
**www.magellangps.com**

NAVMAN, P.O. Box 68155, Newton, Auckland, New Zealand
**www.navmanusa.com**

Northstar Technologies, 30 Sudbury Road, Acton, MA 01720
**www.northstarcmc.com**

RayMarine, Inc., 22 Cotton Road, Unit D, Nashua, NH 03063
**www.raymarine.com**

Simrad, Inc., 1921933 Avenue W, Suite A, Lynnwood, WA 98036
**www.simradusa.com**

Standard Horizon, 10900 Walker Street, Cypress, CA 90630
**www.standardhorizon.com**

# GOVERNMENT RESOURCES

This listing of government resources represents the respective department Web sites for information on the GPS. The FAA Web site is a source for information on WAAS. The US Air Force, Navy and Coast Guard sites are the best government Web sites for basic information about GPS, including information about the satellites.

Aerospace Corporation
**www.aero.org/education/primers/gps/index.html**

FAA
**gps.faa.gov**

National Oceanic and Atmospheric Administration (NOAA)
**www.noaa.gov**

NOAA, Online Charts
**chartmaker.ncd.noaa.gov/mcd/enc/download.htm**

Office of Coast Survey NOAA
**chartmaker.ncd.noaa.gov/**
Smithsonian
**www.nasm.si.edu/exhibitions/gps/**

US Coast Guard
**www.uscg.mil**

US Coastline Extractor
**rimmer.ngdc.noaa.gov/coast/**

US Air Force GPS, Joint Program Office
**gps.losangeles.af.mil/**

USCG Navigation Center
**www.navcen.uscg.gov/**

US Navy
**tycho.usno.navy.mil/gpsinfo.html**

# CARTOGRAPHY SOFTWARE SOURCES

C-Map
**www.c-map.com**

Fugawi
**www.fugawi.com**

Garmin
**www.garmin.com**

GPSS
**www.gpss.tripoduk.com/**

Jeppesen
**www.jeppesen.com**

Maptech
**www.maptech.com**

Navionics
**www.navionics.com**

Nobeltec
**www.nobeltec.com**

RayMarine
**www.raymarine.com**

TopoGrafix
**www.topografix.com**

# A

# Appendix

## PARTIAL LIST OF APRS SYMBOLS AND DESCRIPTION

A data string that makes up a standard APRS position report looks something like this:

**!3612.34N/11518.95W>**

The latitude and longitude are expressed in degrees, minutes and decimal fractions of minutes. This is the standard NMEA format for lat/long output by GPS receivers, and is also the default format for APRS. Thus, the example above says, "36 degrees 12.34 minutes north latitude" and "115 degrees 18.95 minutes west longitude". The character after the longitude, at the end of the string, specifies the symbol that will appear on the monitor screen at the receiving station. In this example, it would be a car.

| Symbol | Description |
|---|---|
| ! | Police or Sheriff |
| # | Digipeater (Green Hollow Star) |
| $ | Telephone |
| & | Gateway |
| ( | Cloudy |
| * | Snowmobile |
| + | Red Cross |
| , | Boy Scouts |
| - | House, QTH with vertical antenna |
| 0 | Circle (Numbered) |
| 1 | Circle (Numbered) |
| 2 | Circle (Numbered) |
| 3 | Circle (Numbered) |
| 4 | Circle Numbered |
| 5 | Circle (Numbered) |
| 6 | Circle (Numbered) |
| 7 | Circle (Numbered) |
| 8 | Circle (Numbered) |

| Symbol | Description |
|--------|-------------|
| 9 | Circle (Numbered) |
| : | Fire |
| ; | Campground, Tent, Portable |
| < | Motorcycle |
| = | Railroad Engine |
| > | Car |
| ? | File Server, Position Server |
| @? | Hurricane, Tropical Storm |
| A | Aid Station |
| C | Canoe |
| E | Eyeball |
| G | Grid Square (6-Digit) |
| H | Hotel (Blue Dot) |
| I | TCP |
| K | School |
| M | MacAPRS |
| N | NTS Station |
| O | Balloon |
| P | Police Car |
| R | Recreational Vehicle |
| S | Space Shuttle |
| T | SSTV |
| U | Bus |
| V | ATV |
| W | National Weather Service |
| X | Helicopter |
| Y | Yacht, Sailboat |
| Z | WinAPRS |
| [ | Runner, Jogger |
| \ | Triangle (Direction Finding) |
| ] | WinLink PBBS (Mailbox) |
| _ | Weather Station |
| a | Ambulance |
| b | Bicycle |
| d | Fire Department |
| e | Horse |
| f | Fire Truck |
| g | Glider, Hang Glider |
| h | Hospital |
| i | Islands On The Air (IOTA) |
| j | Jeep |
| k | Truck |

| Symbol | Description |
| --- | --- |
| m | MIC-Encoder Repeater |
| n | Node |
| o | Emergency Operations Center |
| p | Rover, Dog |
| q | Grid Square (4 Digit) |
| r | Antenna |
| s | Power Boat |
| t | Truck Stop |
| u | Truck, 18 Wheeler |
| v | Van |
| w | Water Station |
| x | X-APRS (UNIX APRS) |
| y | House, QTH with Yagi Antenna |

# B

# Appendix

## USING A COMPUTER SOUND CARD AS A TNC

Most computer sound cards can be used to send and receive packet data; however, they require a TNC management utility to enable them to encode and decode packet tones. The best utility available to have your computer's sound card simulate a TNC is *AGWPE* which is an acronym that stands for AGW's Packet Engine. This utility was written by George Rossopoulos, SV2AGW, and it is free utility you can download at **www.elcom.gr/sv2agw/**

Other than the free utility program the only other thing you will need to turn your sound card into a TNC is a sound card interface, which is a set of cables to connect your sound card to your radio. You can make up these cables or you can purchase them preassembled or as a kit and assemble them yourself.

*AGWPE* will allow you to run baud rates of 300, 1200, 2400 and 9600, and allow you to use the stereo features of your sound card to connect to two radios on two different frequencies. You can also install additional sound cards that can be exclusively used only for packet radio while the sound card that came with your computer can be used for other sound producing applications such as your DVD or CD player.

The *AGWPE* software will run on *Windows 95, 98, ME, XP,* and *2000.* It will not run on *Windows 3.1* or *NT 4.0* or a *DOS* operating system. It also will not work with *UNIX* or Mac programs. It has been successfully installed and run on as basic a computer as the 486/66. *AGWPE* will work with most 16- or 32-bit sound cards, although there are some sound cards like the older ISA cards that are not full duplex capable and may not work with *AGWPE*. The SoundBlaster PCI128 sound cards also have problems with the stereo channels when using *AGWPE* and you can only use them in *AGWPE*'s single port mode.

There are several versions of APRS application software that will work with *AGWPE*, such as *APRSPoint, APRSPlus, OziAPRS, WinAPRS, UI-View,* and *XASTIR.* In Chapter 7 there is a list of Web site addresses where you can download the application software you would like to run.

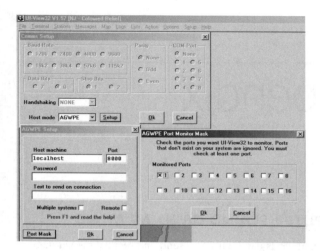

*An example of the AGWPE set-up screens.*

If you don't want to build your own computer sound card to radio interface you either buy a kit with all the necessary parts and assemble the interface cables, or you can purchase the cables already preassembled. Buck Rogers, K4ABT, sells a kit called the *RASCALGLX* which has all the cables and parts needed for a well designed interface. These kits come with isolation transformers for each cable which isolate the computer ground and the radio ground to prevent interference and reduce the risk of stray voltages damaging either the computer or the radio.

Preassembled sound card interfaces are available from Tigertronics, West Mountain Radio, MFJ and *MixW* RigExpert. The Tigertronics interface comes with an auto PTT circuit that eliminates the need for a COM or LPT port for a PTT activation signal. This interface does require external power from 6.75 to 15 Vdc at 10 mA. The West Mountain Radio's RigBlaster Std, Plus, and Pro models allow you to keep both your microphone and sound card attached to the radio's microphone jack. MFJ has two preassembled models: the 1275 and the 1275M. The MixW RigExpert models use a USB transceiver interface.

If you wish to build your own interface, Ralph Milnes, KC2RLM, has a great tutorial on his Web site at **www.soundcardpacket/info**. To get your sound card to handle packet data you only need to download and install the *AGWPE* utility, buy or build a sound card to radio interface, install the APRS or client applications software, and configure your *Windows* operating system. To learn more about using computer sound cards to simulate a TNC visit Ralph's web site or George Rossopoulos', SV2AGW, Web site at **www.elcom.gr/sv2agw/**.

# C

# Appendix

## AN INEXPENSIVE EXTERNAL GPS ANTENNA

*If you operate APRS or just need an external antenna for your GPS receiver, here's one that is easy to build yet offers surprisingly good performance in a compact size. Best of all, it uses commonly available components and materials.*

By Mark Kesauer, N7KKQ, October 2002 QST

This antenna design is based on a classic turnstile configuration (for circular polarization)—two dipoles are placed on the same plane but rotated 90° from each other. These dipoles are then spaced ¼ wavelength above a ground plane. A ¼ wavelength "parallel-plate" transmission line (printed circuit-board material) serves as the connection method and mounting post for the dipoles.

## Construction

Start with the base plate. Cut a 4-inch diameter circle out of thin hobby tin or brass. (It happens that the inside diameter of the container lid is 4 inches, approximately the same width as the hobby tin/brass sheet.) Mark the exact center of the base plate. This is where the parallel-plate transmission line assembly is attached (see Figure 1).

Cut two 4-inch lengths of #14 solid copper or brass wire and bend each in the exact center at 90°.

Figure 1—Close-up view of the coax connection to PCB transmission line and support.

Make the radius of the bend as small as possible. Set these aside, they will be soldered to the parallel-plate section later.

Select an 8-foot length of RG-58/U, RG-174 or RG-188 coax. Attach a male BNC connector to one end (or whatever compatible connector is used on your particular GPS receiver). I used a solderless connector but removed the screw and then soldered the center conductor directly into the screw hole. If your GPS unit has a BNC antenna connection, you can use an Ethernet coax cable found at most computer stores. Just make sure they are 50 Ω. They'll already have the BNC connectors crimped on each end. Just cut in the center, trim to length and you'll have enough for two antennas. The GPS frequency is 1.57542 GHz so the longer the coax, the greater the loss. Use no more than 8

---

## Table 1

### Materials

Hobby tin (K&S #254) or brass sheet (K&S #251) (0.010 thick).
Sheet of single-sided, glass-epoxy PCB material (FR-4 or G10 .062" thick, enough to make two 2-inch pieces 0.250" wide.
Solderless right-angle male BNC connector (RadioShack 278-126) or the appropriate type for your GPS receiver.
8-foot RG-58/U (Radio Shack 278-1314), RG-174 or RG-188 coax.
8-inch #14 solid bare copper or brass wire.
Empty 8-oz cream cheese container.
Misc—Clear 5-minute epoxy or superglue.
Clear spray lacquer, #600 fine sandpaper.

K&S Engineering, 6917 W 59 St, Chicago, IL 60638; voice 773-586-8503; fax 773-586-8556; **www.ksmetals.com/**.
RadioShack, **www.radioshack.com/**.

---

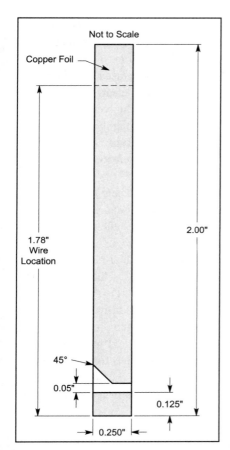

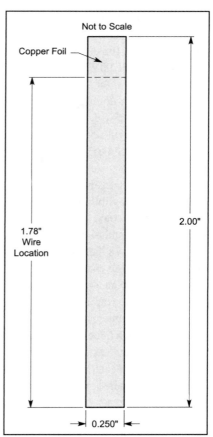

*Figure 2—The active side of the transmission line.*

*Figure 3—The ground side of the transmission line.*

feet—less if you don't need the length.

To make the parallel-plate transmission line, cut two 2-inch lengths of single-sided printed circuit board material that are 0.250-inch wide. Make sure it is glass-epoxy (FR-4 or G10 type material) and that it is 0.062-inch (1/16 inch) thick.

On one of the PCB strips, cut the copper foil with a sharp hobby knife or Dremel tool, as shown in Figure 1. This will be the "active" section of the parallel-plate where the other non-modified strip will be the "ground" side, as shown in Figure 2. The 45° cut on the active side is known as a "microwave turn" which allows the signal to effectively turn 90° to the coax. Glue the two strips together (copper outside) and set aside to dry.

I've found it easier to cut the PCB strips a bit wide and glue them together first. Then I just file both edges to the correct dimensions. A light sanding with #600 sandpaper finishes off the edges and removes any burrs.

Double-sided 0.125-inch thick PCB material could be used but can be difficult to obtain for the average hobbyist. Conversely, by using a single 0.063-inch thick double-sided material we would be working with a rather small and fragile structure (half the thickness equates to roughly half the width). This might not hold up during handling and operation. By using the two sections glued together, we've solved the problem by creating our own 0.125-inch thick material.

Solder the transmission line section to the base plate keeping it as square and plumb as possible. Drill or melt a hole in the plastic container the same diameter as the coax. Feed the end of the coax through the hole and attach the coax to the transmission line active side as shown in Figure 4.

Measure 1.78 inches up from the base end of the parallel-plate section and scribe a line in the copper foil. Solder one of the #14 wires to the ground side of the parallel-plate section. Position as shown in Figure 4. Do the same with the active side—you may need a helping third hand as it's difficult to hold the soldering iron, antenna and position the wires all at the same time.

Measure each leg of the horizontal wires and trim to 1.51 inches from the center junctions. Next, trim both the 45° wires to 1.82 inches from the center junction. If all went well, you should have approximately ½ inch between the tips of the 45° wires and the base. If not, carefully resolder or bend the wires to this dimension.

Using a fine saw or a Dremel tool, remove the excess length of the transmission line just above the wire junctions. Sand the exposed junction to remove any burrs and check for a short circuit.

Note that we've purposely kept the transmission line section length long, until after construction. The thin copper foil tends to separate from the glass epoxy during heavy duty soldering. The longer length acts as a heatsink to preserve the bond between the copper foil and the glass-epoxy base.

## Final Assembly

I've found that an empty, upturned 8-ounce cream cheese container makes a practical radome for the antenna. More importantly, it helps protect the internal workings from mechanical damage.

I usually don't paint the container but I do remove the silk-screened label by using an automotive rubbing compound. It takes some effort but it does come off. Just follow the manufacturer's instructions. Be careful not to apply too much pressure to the lid when you rub the label off. It's made of a different plastic than the container and stretches easily.

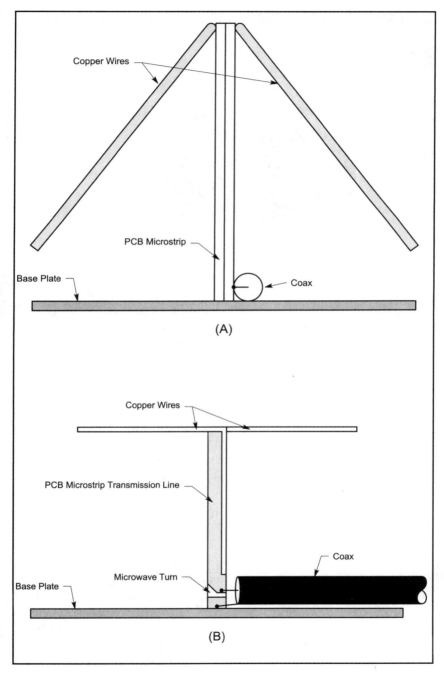

*Figure 4—Side (A) and front (B) views of the parallel-plate transmission line and radiating elements.*

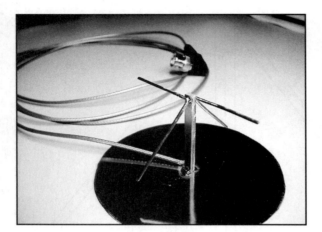

*Figure 5—View of the parallel-plate support with elements attached.*

You should be able to snap the base plate into the lid of the container. It's a tight fit so just work your way around the lid until the entire base plate is flush with the lid bottom You might have to cut a notch in the lip of the lid to allow the coax to exit the unit cleanly. Carefully align the coax with the lid notch and snap the cover onto the lid. It's normal for the top of the transmission line assembly to slightly raise the "bump" on the container bottom.

## Theory of Operation

In a normal turnstile, we would have a double dipole configuration with both dipoles on the same plane but rotated 90° from each other. Additionally, the second dipole is fed 90° out of phase with another ¼ wavelength of coaxial cable (see Notes 2 and 3). This creates some difficult assembly problems since you would have to isolate the second dipole section from ground while maintaining the tight distance and spacing requirements. Due to the size constraints, this second dipole connection would require a very small diameter coax that might be difficult to work with and even harder to obtain. With this antenna, we cheat a bit and use a self-phased quadrature type feed.

To obtain circular polarization without a coaxial phasing line, the shorter dipole is cut so its impedance is $50 - j50\ \Omega$. The longer dipole is fashioned into an inverted V shape and cut so its impedance is lowered to $50 + j50\ \Omega$. With the combined asymmetrical dipoles and with them spaced slightly closer than ¼ wavelength to the ground plane, the antenna's impedance is near 50 Ω with a much more omnidirectional pattern, an important consideration for reception of GPS satellites close to the horizon.

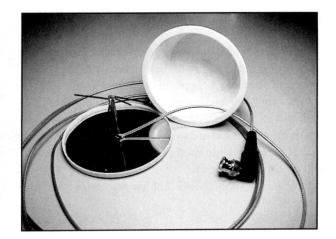

Figure 6—The
finished GPS
antenna with
radome ready to
snap into place.

## Operation

Connect the antenna to the GPS receiver and watch the signal-strength
indicator. You should see an improvement over the supplied stock antenna. You
can tweak the antenna by bending the wires up and down gently and watching
the results on your GPS unit. Be careful of the solder joint—it's rather fragile.
Adjust for maximum displayed signal. Repositioning the antenna may also
improve reception. With this antenna, I routinely receive five to eight satellites
on my Garmin II receiver.

If you are using a GPS unit that sends dc voltage up the coax to power
an external preamp or amplified antenna, don't worry. Since the elements are
not grounded or shorted, there is no dc path. Just be careful not to let either
end of the active elements touch ground. [Be advised that some GPS receivers
with internal patch antennas have an antenna switching circuit. This circuit
disables the internal antenna when an amplified external antenna is attached.
The receiver senses current flow that is intended to power the amplifier of the
external antenna. If your receiver has this feature, you will want the switch to
activate and disable the internal patch antenna. Placing 1 kΩ to 5 kΩ across the
ground and center conductor of the coax should be sufficient. Check with the
manufacturer of your GPS receiver.—*Ed.*]

If you are mobile, most GPS receivers will do a fair job of receiving
signals through the windshield of a car. During the summer, however (and
especially out here in the Southwest), the GPS gets baked while sitting in the
hot sun. Obviously, one way to solve this problem is to locate the receiver
somewhere cooler and place this external antenna on the dash. Once that's

done, you've protected your expensive receiver while sacrificing a $10 antenna.

This design was not intended for outside use, hence the lack of external mounting suggestions. Any prolonged exposure to the elements will degrade the antenna's unprotected metal parts. My suggestion—if you plan on using this unit outside, at least spray the metal parts with clear lacquer and seal the exposed coax end with RTV. Otherwise, don't waste a lot of time weatherproofing it. Because these antennas are so cheap and easy to build, if one does deteriorate throw it away and build another. Perhaps you might want to keep a couple of spares on hand, just in case.

My thanks to Zack Lau, W1VT, of the ARRL Lab for his advice and expertise.

### Notes

[1]Tom Hill, WA3RMX, "A Triband Microwave Dish Feed," *QST*, Aug 1990.
[2]Zack Lau, W1VT, "A Simple 10-Meter Satellite Turnstile Antenna (RF)," *QEX*, Nov 2001.
[3]Zack Lau, W1VT,"A Simple 10-Meter Satellite Turnstile Antenna (Feedback)," *QEX*, Jan 2002.

*Mark Kesauer, N7KKQ, received his Novice license in 1969. He has held the call signs WN8CGM, KA5ZCH and his present Extra class call sign. Mark belongs to QCWA, Ten-Ten International and has been a longtime ARRL Member. Mark holds an Engineering degree and has been a Computer Aided Designer (printed-circuit boards) for 29 years. Mark may be contacted at* **n7kkq@arrl.net.**

# Index

# Notes

# Notes

# Notes

# Notes

# Notes

# Notes

# Notes

# Notes

# FEEDBACK

Please use this form to give us your comments on this book and what you'd like to see in future editions, or e-mail us at **pubsfdbk@arrl.org** (publications feedback). If you use e-mail, please include your name, call, e-mail address and the book title, edition and printing in the body of your message.
Also indicate whether or not you are an ARRL member.

Where did you purchase this book?
  ☐ From ARRL directly        ☐ From an ARRL dealer

Is there a dealer who carries ARRL publications within:
  ☐ 5 miles      ☐ 15 miles      ☐ 30 miles   of your location?      ☐ Not sure.

License class:
  ☐ Novice ☐ Technician ☐ Technician Plus ☐ General ☐ Advanced ☐ Extra

Name_____ ARRL member?  ☐ Yes ☐ No
_____ Call Sign _____
Daytime Phone   (     ) _____ Age _____
Address _____
City, State/Province, ZIP/Postal Code _____
e-mail address_____
If licensed, how long? _____
Other hobbies _____
Occupation _____

| For ARRL use only | GPS |
|---|---|
| Edition | 1 2 3 4 5 6 7 8 9 10 |
| Printing | 1 2 3 4 5 6 7 8 9 10 |

From _____

_____

_____

EDITOR, GPS AND AMATEUR RADIO
AMERICAN RADIO RELAY LEAGUE
225 MAIN STREET
NEWINGTON CT 06111-1494

——————————————— please fold and tape ———————————————